Adaptive Data Compression

THE KLUWER INTERNATIONAL SERIES
IN ENGINEERING AND COMPUTER SCIENCE

COMMUNICATIONS AND INFORMATION THEORY

Consulting Editor:
Robert Gallager

Other books in the series:

Digital Communication. Edward A. Lee, David G. Messerschmitt
ISBN: 0-89838-274-2

An Introduction to Cryptology. Henk C.A. van Tilborg
ISBN: 0-89838-271-8

Finite Fields for Computer Scientists and Engineers. Robert J. McEliece
ISBN: 0-89838-191-6

An Introduction to Error Correcting Codes With Applications.
Scott A. Vanstone and Paul C. van Oorschot
ISBN: 0-7923-9017-2

Source Coding Theory. Robert M. Gray
ISBN: 0-7923-9048-2

Switching and Traffic Theory for Integrated Broadband Networks.
Joseph Y Hui
ISBN: 0-7923-9061-X

ADAPTIVE DATA COMPRESSION

by

Ross N. Williams
Australian National

foreword by

Glen G. Langdon, Jr.
University of California, Santa Cruz

KLUWER ACADEMIC PUBLISHERS
Boston/Dordrecht/London

Distributors for North America:
Kluwer Academic Publishers
101 Philip Drive
Assinippi Park
Norwell, Massachusetts 02061 USA

Distributors for all other countries:
Kluwer Academic Publishers Group
Distribution Centre
Post Office Box 322
3300 AH Dordrecht, THE NETHERLANDS

Library of Congress Cataloging-in-Publication Data

Williams, Ross N. (Ross Neil), 1962–
    Adaptive data compression / by Ross N. Williams; foreword by Glen
C. Langdon, Jr.
      p.   cm. — (The Kluwer international series in engineering and
computer science. Communications and information theory)
    "A reproduction of a a Ph. D. thesis that was submitted to the Dept.
of Computer Science at the University of Adelaide on 30 June 1989"-
-Acknowl.
    Includes bibliographical references (p.   ) and index.
    ISBN 0–7923–9085–7
    1. Data compression (Telecommunications) I. Title. II. Series.
TK5102.5.W539 1991
005.74'6—dc20   90-5059
                                                                   CIP

Contents

List of Figures

List of Tables

Foreword

Following an exchange of correspondence, I met Ross in Adelaide in June 1988. I was approached by the University of Adelaide about being an external examiner for this dissertation and willingly agreed. Upon receiving a copy of this work, what struck me most was the scholarship with which Ross approaches and advances this relatively new field of adaptive data compression. This scholarship, coupled with the ability to express himself clearly using figures, tables, and incisive prose, demanded that Ross's dissertation be given a wider audience. And so this thesis was brought to the attention of Kluwer.

The modern data compression paradigm furthered by this work is based upon the separation of adaptive context modelling, adaptive statistics, and arithmetic coding. This work offers the most complete bibliography on this subject I am aware of. It provides an excellent and lucid review of the field, and should be equally as beneficial to newcomers as to those of us already in the field.

A paper "Modeling and Coding" by Rissanen and Langdon lays some theoretical groundwork to adaptivity, and Ross extends this work by further categorizing the notion of causal adaptation. In so doing, he provides insights into possible implementations. Ross's introduction also comments on terminology problems. For example, the term "compression ratio" has no standard meaning between authors. Instead of arguing what compression ratio "really means", Ross proposes new terms that imply how the compression performance is calculated. For example, if value x is identified as the "proportion remaining", the reader immediately knows that the number of bytes in the compressed file is x times the number of bytes in the uncompressed file.

The capability to adapt to the behaviour of a zero-order Markov source opens many doors, since a variable-order Markov source is a collection of independent zero-order sources. The arithmetic coding component of the modern compression paradigm employs relative frequencies directly as opposed to older techniques that must first map the frequencies onto sets of integer-length code words possessing the prefix property. Ross points out the historical interest of an algorithm called DAFC (Double Adaptive File Compression) that first adapts to Markov contexts and then adapts a probability distribution for each context. Ross's algorithm of Chapter 2, DHPC (Dynamic History Predictive Compression), does the kind of double adaptation that DAFC would have liked to have done. DHPC, Ross's opening contribution and the basis for the three following chapters, is a valuable original contribution in its own right. A backwards tree of maximum depth m represents the more frequent variable-order contexts up to Markov order m. The algorithm description is generic, and can be used for experimentation. Designer-selected parameters determine if a node in the tree has enough activity to serve as a context, or enough activity to spawn deeper nodes.

The DHPC algorithm leads to the Swiss Army Knife Data Compression (SAKDC) algorithm. SAKDC accommodates, through parameter selection, the DHPC algorithm as a special case. Ross defines different forms and mechanisms for adaptivity of variable-order models, and SAKDC provides an implementation environment for experimenting with them. The experiments performed provide a page of conclusions regarding what seems to work and what doesn't work in variable-order Markov models.

A very interesting aspect of the work is the identification and investigation of multimodal sources. A cylinder of a disk or a magnetic tape of data may contain Pascal programs, hex dumps, and text information. These three types of data each correspond to a source mode. For each source mode, the compression algorithm should employ a model appropriate for that mode. Multi-Modal algorithms switch between models as the source jumps from one mode to another. The MMDC (Multi-Modal Data Compression) algorithm is based upon the above approach, and an implementation satisfactorily switches between models. Ross observes that MMDC illustrates science building upon itself; MMDC unifies many concepts from the modern paradigm of data compression that were developed without multi-modal compression in mind.

Glen G. Langdon, Jr.
Computer Engineering Department
University of California, Santa Cruz
June 1990

Preface

This book is identical to my Ph.D. thesis (submitted to the University of Adelaide on 30 June 1989) except that it has been re-typeset and a few minor corrections have been made. These are listed in the presentation notes.

The thesis reviews the field of text data compression (Chapter 1) and then addresses the problem of compressing rapidly changing data streams (Chapter 2–7). To compress such streams, the compressor must continuously revise its model of the data (Chapter 2–4), and sometimes even discard its model in favour of a new one (Chapter 5).

Although the techniques involved in data compression are often complicated, the ideas are simple, and throughout the thesis I have endeavoured to keep them so. The reader should be able to tackle the thesis with no knowledge of data compression.

Ross N. Williams
June 1990
Adelaide, Australia

Acknowledgements

This monograph is a reproduction of a Ph.D. thesis that was submitted to the Department of Computer Science at the University of Adelaide on 30 June 1989 and I am grateful to all those who supported me during my candidature. In particular I am very grateful to my supervisor, Dr. William Beaumont for his guidance and support. I am also grateful to my temporary supervisor Dr. Tao Li for his help while Dr. Beaumont was on study leave in 1987.

The research that led to the thesis was supported in part by a Commonwealth of Australia Postgraduate Research Award.

Thanks is recorded to those whose informal communications were quoted in the thesis, as all have given additional permission to be quoted in this book.

Presentation Notes

References: All references are cited in the form [*<firstauthor><year>*]. Citations are set in bold face the first time they appear and ordinary face thereafter. All references cited in the text appear in the reference list and the index. Citations of references in the same year by the same author appear identical.

Special terms: New or important terminology has been set in bold face and appears in the index.

Tables and Figures: Unless otherwise stated, all tables and diagrams appearing in this thesis were created specifically for the thesis and are original to the thesis. All references to tables and figures in the text are set in bold face so that the text describing any particular table or figure can be found quickly.

Captions: Every table and figure in the thesis has a title which summarizes the diagram and appears in the table of contents. In addition, some tables and figures have a caption (a paragraph of descriptive text). Captions have been included mainly to aid the casual reader.

Intra-document references: References to entities within this thesis appear with the first letter in upper case (e.g. "Section 78.5"). References to entities in other documents appear with the first letter in lower case (e.g. "section 78.5").

Italics: *Italics* is used both for emphasis and for program and algorithm identifiers.

Notation: Section 1.2 describes much of the notation and terminology of the thesis. Appendix E contains a summary of the mathematical notation used in this thesis.

Typesetting: This book was typeset using the TEX[**Knuth79**][**Knuth84**] typesetting system and was reproduced photographically by Kluwer Academic Press.

Graphics: Most of the diagrams of this thesis were generated on a *Macintosh*[1] computer using the program *MacDraw*.[2] Histograms were produced on a *Macintosh* using the program *Excel*.[3] The tree diagrams of Chapter 2 were plotted from *PostScript*[4] code generated directly by a DHPC program. The graphs in Chapter 4 and Chapter 5 were generated by a plotting program developed by the author. The "range frame" style of these graphs was inspired by the graphic design book [**Tufte83**] (esp. p. 132).

Corrections

This book is identical to my Ph.D. thesis except that it has been re-typeset and corrected. The corrections are:

Front matter: The front matter was reworked to suit the form of a book, and the foreword and preface were added.

Chapter 1: A note was appended to footnote 18 (Section 1.3.1) and the word "bits" in the same footnote was set in boldface and entered into the index. A copyright notice was added to **Figure 10**.

Chapter 4: The title of **Table 39** was corrected.

Conclusions: A closing parenthesis was inserted in the fifth primary contribution paragraph.

Appendix A: Reference to Jones in the second paragraph was made less ambiguous. Typographical errors in the second and last equations of the proof were corrected.

Appendix C: "two pages" was changed to "few pages".

Appendix E: The definition of M was altered.

Detected remaining errors: The word "wooloomooloo" used throughout the thesis as an example message is a misspelling of the famous suburb of Sydney, Australia. The correct spelling is "Woolloomooloo". Enthusiastic readers way wish, as an exercise, to re-work all the examples using the correct spelling.

---

[1] *Macintosh* is a trademark of Apple Computer, Inc.

[2] *MacDraw* is a trademark of Claris Corporation.

[3] *Excel* is a trademark of Microsoft Corporation.

[4] *PostScript* is a trademark of Adobe Systems Incorporated.

Abstract

The class of Markov data compression algorithms provides the best known compression for ordinary symbol data. Of these, the PPM (Prediction by Partial Matching) algorithm by Cleary and Witten (1984) is the most successful. This thesis reviews the class of Markov algorithms and introduces algorithms similar to PPM which are used as a vehicle for examining adaptivity in data compression algorithms in general.

Despite the amount of research into adaptive data compression algorithms, the term "adaptive" is not well defined. The term can be clarified by viewing the problem of adapting as that of tracking a source through a source space using the data the source is generating as a trail. Adaptive algorithms achieve this by making assumptions about source trajectories. These assumptions can be used to classify data compression algorithms into four groups with respect to adaptivity: not adaptive, initially adaptive, asymptotically adaptive and locally adaptive. These groups correspond roughly to classes of source trajectory. A new class of source called multimodal sources is introduced, members of which jump among a finite number of distinct points in the source space.

After common source trajectories have been identified, modifications to Markov algorithms are described that enable them to track each kind of source. Some of the modifications involve complex algorithms which are discussed in detail. Experimental results show that these modifications can substantially improve compression performance.

Of particular interest are sources with a multimodal trajectory. Such sources can be compressed well using a combination of locally adaptive and asymptotically adaptive models. The result is a multimodal algorithm which is independent of any particular sub-model. Experimental results support this approach.

Finally, there is a discussion of the possible applications of data compression techniques in the field of user interfaces. By predicting the user's behavior and presenting the predictions to the user in an invokable form, it is possible to eliminate redundancy in the user interface just as it can be eliminated in data streams.

Adaptive Data Compression

Chapter 1

Introductory Survey

1.1 Introduction

The purpose of data compression is to remove redundancy from data so that it takes less time to transmit and less space to store. Data compression increases system throughput, improves network security and relieves programmers of the task of packing data efficiently. Because data compression operates at the logical level, it cannot be made obsolete by advances in storage or network technology.

The purpose of data compression research is to develop and analyse methods for representing information in the minimum amount of space. Data compression research is important to Computer Science because it explores one extreme of the compactness/speed trade-off present in *all* decisions about data representation. To store or transmit data, a representation must be *chosen*; there are no default representations. With this in mind, we proceed in the examination of data representations with the one-eyed view that compression is paramount.

1.1.1 Compression as Representation

In order to be stored or transmitted, information must first be represented in a physical medium. Choosing a representation means associating meaning to the states of a physical object. Representations vary in two respects:

- the amount of physical medium used.

- the processing power needed to perform various operations.

Unfortunately, representations that minimize one cost do not always minimize the other and a compromise is usually necessary.

We view data compression as the translation of a message from one representation to a different representation that uses less space. Data compression is defined *relatively*, and strictly it is incorrect to refer to any particular representation as "compressed" without first defining an "uncompressed" representation to which it can be compared. In practice, compression performance is usually expressed relative to an unstated set of **natural representations** which is the set of minimum-length representations that use the same amount of space for all data values; "compressed" representations use less space *on average*.

In order to use less space than a natural representation, compressed representations use code strings of varying length to represent the different data values. This variation in length often increases the time required to perform operations on compressed representations. For example, elements of a compressed array can no longer be accessed at random. Thus, we expect to find data compression only where the cost of space is high relative to the cost of time.

1.1.2 Data Compression in Ancient Greece

The problem of choosing a representation has been around for as long as humanity has had to represent information. In ancient Greece, the cost of papyrus was orders of magnitude more expensive than the cost of paper is to us today.[5] As a result, texts were written with no punctuation and no spaces, yielding a space saving at a cost of reading time. The ancient Greeks and Romans used a telegraph system that employed varying numbers of torches to convey letters of the alphabet [Havelock78](pp. 86–87) but it employed neither data compression nor error correction.

In the late eighteenth century, the British Admiralty employed a series of cabins that transmitted six-bit signals using shutters.[6] Some of the $2^6 = 64$ permutations were mapped onto the alphabet and the remaining code space was allocated to common words and phrases, resulting in a form of data compression. In particular, one code, used

---

[5] Dr. R.F. Newbold of the University of Adelaide Classics Department has estimated it as 100 to 1000 times more expensive (private communication (permission to quote kindly granted 12 May 1989)).

[6] The shutter cabin story appears in the introductory chapters of a book "Text Compression"[Bell89] (to appear) and I am grateful to the authors for discovering the story. The book contains further details of the system and some references.

to represent the message "Sentence of court martial to be put into execution", simultaneously provided a high degree of compression and a "fatal" vulnerability to errors. More serious omissions of error detection have occurred in our own time. The worst case was probably that of a nuclear alert being triggered by the failure of a ten cent communications chip (a 74175). The chip caused the "number of missiles" field in an unchecksummed network filler packet to go positive and signal a major attack[**Borning87**](p. 113).

In recent years, the acceleration of technology has resulted in some long strings of words falling into common use. The response has been to replace them with newly created shorter words. The new word is typically an acronym. For example the phrase "Central Processing Unit" is now usually written as "CPU". Acronyms have been used as a data compression technique for ages. In Roman civilization, tombstone space was costly as was the cost of engraving. The result was the common S.T.L. (Sit Terra Levis) which means "Let the earth rest lightly upon her".[7] Acronyms were also used by the Romans to squeeze lengthy imperial honours onto coins.

Perhaps the earliest widely-known example of data compression in a representation is Morse code which makes some attempt to allocate shorter codes to the more frequent letters. In the normally used variant of Morse code, the letter Y (— · — —) takes three and a half times as long to transmit as the letter E (·).

The result of the drive to reduce sentence length can be found in many languages whose commonly used words are shorter than rarely used words. In general, the more often a word is used, the shorter it is [**Zipf49**](p. 64). Some common combinations, such as "do not" ("don't") are explicitly shortened using an apostrophe.

Another example of compression can be found in the Arabic number system which represents numbers by digits of increasing powers of a base (using $O(\log_b n)$ space) rather than the simpler system of using a number of strokes ($O(n)$ space).

The natural world is full of examples of data compression. A particularly good example is the coding of genetic information. Ordinary DNA is highly redundant, containing long tracts of unused bases. Viral DNA, on the other hand, is under strong selection pressure to become small. The result is that "some small viruses (like $\phi X174$) [have] evolved

---

[7] Others were D.M. for Dis Manibus meaning "To the ghosts of the underworld" and B.M. for Bene Merenti meaning "To one deserving well".

overlapping genes, in which part of the nucleotide sequence encoding one protein is used (in the same or a different reading frame) to encode a second protein."[Alberts83](pp. 239–240)(emphasis by Alberts). This phenomenon can be compared with the superstring problem discussed in Section 1.5.2.5.

Finally, we find examples of data compression in everyday life. Newspaper headlines are set in a large font which is time efficient (easy to read, even at a distance) but is space inefficient (uses lots of space). On the other hand, the classified advertisements are set in a font which is time inefficient (hard to read) but space efficient.[8] The examples of data compression are summarized in **Table 1**.

Context	Technique
Ancient Greece	Spaces omitted
Shutter Cabin	Spare codes used to transmit phrases
Acronyms	Words replaced by letters
Morse Code	Shorter codes assigned to frequent letters
Arabic Digits	Base b used rather than base 1
Viral DNA	Genes overlap in the nucleotide sequence
Newspapers	Use of fine print

Data compression techniques are not the exclusive domain of computer systems. This table lists some uses of data compression in everyday life.

Table 1: Examples of everyday data compression.

The requirement for simple and fast decodability by humans restricted the field of data compression to representations that have a one-to-one correspondence between objects and their representations. With the advent of computers, representations which would be prohibitively expensive for humans to use, have suddenly become feasible.

1.1.3 Founding Work by Shannon

In 1948 Shannon[Shannon48][9] laid the foundation of the field of information theory, a discipline that concerns itself with the communication of information over noisy channels.

Shannon divided communication into five components (**Figure 1**).

---

[8] Helman and Langdon used the example of legal fine print to make the same point [Helman88].

[9] This paper was later reprinted in a book[Shannon49] that also contained a related paper by Weaver.

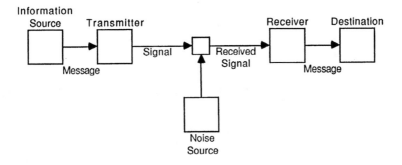

Shannon's model of communication (1948) revolves around an abstracted channel through which a source communicates with a destination. The transmitter and receiver are used to overcome the limitations (e.g. noise) of the channel.

Figure 1: Shannon's model of communication.
(Redrawn from figure 1 of [Shannon48])

1. An **information source** which produces a changing value over time. Shannon's model is fairly general, encompassing multidimensional continuous sources as well as discrete sources.

2. A **transmitter** which translates the message into a signal suitable for transmission over the channel.

3. A **channel** which is the medium used to transmit the signal. The channel can use any physical process as long as it carries the signal.[10] Shannon modelled only the channel's capacity to convey information. Both noiseless and noisy channels are considered.

4. A **receiver** which translates the channel signal into the message.

5. A **destination** which is the process for whom the message is intended.

Shannon's model is general and covers many aspects of communication including error correction, data compression and cryptography.

---

[10] For the purposes of theoretical data compression, disks and networks can both be treated as channels. In practice each can take the role of the other. If user A sends a mail message to user B on the same machine, the disk on which the message is stored acts as a small network. Conversely, a megabyte of data could be stored on a network and retrieved a day or so later by transmitting it to a non-existent network address on the other side of the world.

Although these three aspects are closely related (Section 1.16), it is wise to separate them in theory and practice as they have conflicting design goals. In communication systems that employ all three techniques, messages are typically compressed by a data compression layer, encrypted and then expanded by an error correcting layer. The three transformations are performed in reverse order at the receiving end.

As we are chiefly concerned with data compression, we will operate under the following assumptions which are additional restrictions on Shannon's model:[11]

Assumption 1: Sources produce a sequence of symbols from a fixed, finite source alphabet.

Assumption 2: The channel carries a sequence of symbols from a fixed, finite channel alphabet.

Assumption 3: The cost of transmitting each channel symbol is identical.[12]

Assumption 4: The channel is noiseless.

It is not clear whether it is best to model a source as a random process that produces an infinite stream of symbols (as is implicit in Shannon's approach) or as a random variable which produces a single finite string (as proposed by Rissanen and Langdon[**Rissanen81**]). Both models are useful. Infinite messages can be used to model communication lines and finite messages can be used to model the compression of files. In this thesis both models will be used.

1.2 Common Notation and Data Structures

A number of constructs, concepts and data structures arise so often in data compression that we present them here before continuing. The section starts with notation and moves onto data structures. Discussion of data structure *implementation* is deferred until later sections.

---

[11] For a detailed discussion of more general cases, see Shannon's original paper [Shannon48]. For an additional overview of the whole field of information theory see [Usher84].

[12] Work has been done on channels with unequal symbol costs[**Perl75**][**Altenkamp78**] but like much of this early work, it has largely been superseded by arithmetic coding and in this case a particular version of arithmetic coding described by Guazzo[**Guazzo80**].

Permeating all data compression is the notion of **symbol**. Symbols come in two types. **Source symbols** are used to construct source messages. **Channel symbols** are carried by the channel and are used to communicate the message. In practice the two sets of symbols usually coincide. In this thesis, n is a constant used to denote the number of different source symbols and A is used to denote the fixed, finite, ordered set of n symbols, with $A = \{a_1 \dots a_n\}$. The set of channel symbols is not defined formally, as the focus of the thesis is on source modelling.

Experience has shown that it is important to distinguish[13] between *symbols* and *instances of symbols* which in this thesis are referred to simply as **instances**.[14] The set of symbols defines the set of values that symbol-valued objects (instances) can take. A **message** (also called a message string) consists of a sequence of instances. Instances can be thought of as Petri net[**Peterson77**] like tokens that come in n different colours[**Genrich81**] and flow through data compression systems. The set of symbols can be thought of as the set of colours of such tokens. There are only ever exactly n symbols but the number of instances in existence can vary from zero to infinity.

A **sample** is a bag of instances. Samples are usually represented in the form of a frequency for each symbol with the sum of the frequencies corresponding to the number of instances in the sample.

S is used to denote the set of all finite strings of instances, with S_l denoting the set of all strings of length l instances. ϵ denotes the empty string. The set of all strings of a given length is ordered. If x is a string then x_k denotes the k'th instance in x, with x_1 being the first (leftmost) instance. $|x|$ denotes the length of x. If x and y are strings or instances then xy denotes the concatenation of x and y. $x_{k\dots l}$ denotes $x_k \dots x_l$. A set of strings satisfy the **prefix property** if no string is a prefix of any other string in the set. The term **history** is used to refer to the string consisting of the concatenation of all the instances generated by the source to date, with the last (rightmost) instance of the string being the most recently generated instance. In formal descriptions, h denotes the history string.

R denotes the set of real numbers. **Z** denotes the set of integers. It is convenient, when introducing constants and variables, to introduce their domain as well. Thus the phrase "we introduce a constant $x \in \mathbf{Z}[0,10]$" introduces the integer constant x which lies in the range $[0,10]$. Square

---

[13] As far as the author is aware, this distinction is a new one.

[14] Other names considered were "event", "arrival", "occurence", "element", "item", "observation", "outcome" and "symbol instance".

brackets denote a closed interval ($x \in [a, b] \Rightarrow a \le x \le b$), round brackets an open one ($x \in (a, b) \Rightarrow a < x < b$). The word "iff" is used as an abbreviation for the phrase "if and only if".

The first data structure that arises repeatedly is the **history buffer**. Many data compression algorithms require access to the most recent m instances generated by the source. The history buffer stores these in m slots each of which contains an instance. The slots are numbered $\mathbf{Z}[1, m]$ with slot 1 holding the most recently received instance (the youngest) and slot m holding the least recently received instance (the oldest) (**Figure 2**).

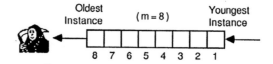

Many data compression techniques require random access to the most recent m instances the source has generated. A sliding history buffer fulfills this need.

Figure 2: A history buffer.

When a new instance arrives, the instances in the history buffer are shifted one slot. The instance in slot i is moved to slot $i+1$. The instance in slot m is discarded and the newly arrived instance is placed in slot 1. History buffers provide the same functionality as fixed-length queues but also allow any of their elements to be read.

The second common data structure is the **digital search tree**, sometimes referred to as a **trie[Knuth73]**(section 6.3). Because "trie" looks like a spelling mistake, we will use "tree" instead. Unless otherwise stated, the word "tree" in this thesis is used to refer to a digital search tree whose arcs are labelled by symbols.

A digital search tree consists of a set of nodes with one node being distinguished as the root node. Each node can have from 0 to n **child nodes**. The term **arc** is used to refer to the link between a node and one of its child nodes. The term **branch** is used to refer to a sequence of arcs and nodes commencing at the root node and connecting nodes of monotonically increasing depth. A branch need not terminate at a leaf node.

Each arc in a tree is labelled with a symbol that is different from those of its sibling arcs. Each node in the tree corresponds to the string constructed by concatenating the symbols labelling the arcs on the path from the root to the node in question. This string is known as the **node's string**. Nodes are most easily referred to by their string. In a **forwards tree**, the symbol labelling the arc connecting to the root node forms the first (leftmost) instance of the string; in a **backwards tree** it forms the last (rightmost) instance (**Figure 3**). The root node always represents the empty string.

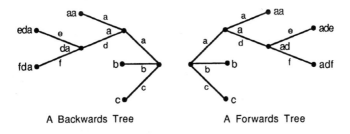

A Backwards Tree   A Forwards Tree

Backwards and forwards digital search trees arise repeatedly in data compression. Each node corresponds to the string constructed by moving from the root to the node. The root node corresponds to the empty string. The only difference between backwards and forwards trees is the direction of their strings.

Figure 3: Backwards and forwards digital search trees.

The depth of a tree node is defined to be the number of arcs linking the node to the root node. The root node has a depth of zero. The depth of a tree is defined to be the depth of the deepest node in the tree. When we refer to the number of nodes in a tree, the root node is included; a tree consisting of only the root node contains one node.

Trees are often augmented with extra attributes. Information is usually attached to the nodes. Sometimes probabilities are associated with each arc.

A tree is **degenerate** iff every node has less than two child nodes. A tree is **balanced** iff the heights of the two subtrees of every node differ by at most 1 (from [**Wirth76**](p. 215)).[15] A tree is **uniformly k-furcated** if every non-leaf has exactly k child nodes. A tree is **uniformly furcated** if it is uniformly-k-furcated for some k. A tree is **solid** if every leaf is of the same depth and the tree is k-furcated where k is the maximum

---

[15] In contrast, the term **unbalanced trees** is used to refer to a method of managing trees that does not attempt to maintain balance.

furcation allowed (usually $k = n$). A tree is **solid to depth d** if it is solid and one of its leaves is of depth d.

Digital search trees arise repeatedly in data compression. A tree can be used to store a dictionary of strings. The set of strings corresponding to the leaves of a tree is guaranteed to satisfy the prefix property.

Trees are often used in conjunction with history buffers. A node in a tree **matches** the history if the node's string x is identical to the string formed by the $|x|$ most recent elements of the history.[16] Thus, for any given history and tree, there is a distinguished branch stretching out from the root such that the string of each node on the branch matches the history. This is called the **matching branch** and the nodes on the branch are called **matching nodes**. The deepest node of a matching branch, which is not necessarily a leaf node, is called the **tip** of the branch. The depth of the tip node is constrained by the length of the history as well as the depth of the relevant part of the tree.

Algorithms in this thesis will be specified in an Ada-like [USDOD83][17] programming language. Liberties with the notation have been taken so as to simplify the code. The **inc** operator increments its integer argument. The **dec** operator decrements its integer argument. The construct

> **loop**
> <*statements*>
> **exit if** <*condition*>;
> <*statements*>
> **end**

describes an infinite loop that is terminated only when control encounters an "**exit if** <condition>" statement whose condition evaluates to *true*. The statement

> **return** <*expression*>

terminates execution of a function, returning the value of the expression as the function's value.

---

[16] With the symbol labelling the arc connected to the root node being compared with the most recent instance in the history in the case of a backwards tree and being compared with the $|x|$'th most recent instance in the history in the case of a forwards tree. A node's string does not match if the history is shorter than the string.

[17] Ada is a registered trade mark of the US-Government-Ada Joint Program Office.

exists(u) is a function that returns *true* iff node u exists. *isleaf*(u) is a function that returns *true* iff u is a leaf. The procedure *new* creates a new node and assigns it to its argument. The new node's attributes are automatically initialized to "sensible" values (e.g. in the case of a sample of instances an empty sample).

Program variables are declared inline with the code. Their scope extends to the end of the enclosing construct.

1.3 The Problem of Data Compression

Shannon's model presents the problem of data compression as that of constructing transmitters and receivers that can translate between compressed and uncompressed representations. It is worth spending some time elaborating upon this problem for, as history has shown, the manner in which the problem is approached radically affects the solutions that are apparent.

1.3.1 Real vs Quantized Information

Shannon devised a measure for the quantity of information H that knowledge of the occurrence of an event of probability p yields.

$$H(p) = -\log p$$

The log is to the base of digit that H is to be measured in. Rather than use indefinite logarithms, we choose base e as a suitable base[18] and use natural logarithms (ln) throughout.

Events that have a high probability convey little information when they occur. Events that are unlikely convey a lot of information. The quantity $H(p)$ can take any real value in the range $[0, \infty]$. Information is a continuous quantity.

---

[18] It is common to refer to digits of base two as **bits** after the contraction binary digits. The author proposes that digits of base e be called **nats** after natural digits (Purists please note: "nits" would be confused with "bits", both lexicographically and phonetically). Base e has the advantage of forcing the user to recognise that information is a continuous quantity, not quantized to bits. Shannon called base e units **natural units**[Shannon48](p. 380). **Note:** Since submission of this thesis, Jones (C.B.) (private communication) has pointed out that the term "nat" appears on page 12 of a book by N. Abramson, "Information Theory and Coding" published by McGraw-Hill, New York, in 1963. The book gives "nat" as a contraction of "natural unit", "bit" as a contraction of "binary unit", and distinguishes between information units (such as bits (for binary units)) and storage units (such as **binits** (for binary digits)).

In contrast, information in the digital world of computers is quantized. A memory unit can be in one of $n \in \mathbf{Z}[2, \infty)$ states. In practice n is usually a power of two. The amount of information held by the unit is

$$-\sum_{i=1}^{n} p_i \ln p_i$$

where p_i is the probability of the unit taking the value i. This is maximized when each p_i is $1/n$ in which case the register always holds exactly $\ln n$ nats; to assume any other probabilities would be wasting space. Thus the set of all quantities of information that can be stored as independent units in a computer is

$$\{\ln i \mid i \in N\},$$

a set of real values which we will call the **bucket set**. Each **bucket** in the set has a **bucket size** i and a **bucket capacity** $\ln i$.

The problem of representing continuous sized pieces of information using buckets of quantized size has been central to the difficulties faced by the field over the last forty years. The traditional solution to the problem is the technique of blocking.

1.3.2 Pure Blocking

Consider the problem of converting a stream of numbers of base m into a stream of numbers of base n with $n \geq m$. The simplest method, of copying each input bucket into each output bucket, wastes a proportion

$$w(m, n) = \frac{\ln n - \ln m}{\ln n}$$

of the space in the output buckets. Efficient conversion can be achieved by buffering a input buckets and b output buckets, effectively forming two large buckets of size n^a and m^b. The closer the sizes of these buckets, the more efficient the conversion. The values of a and b that are most efficient can be determined by minimizing

$$w(a, b) = \frac{\ln n^b - \ln m^a}{\ln n^b} = 1 - \frac{a \ln m}{b \ln n}$$

which is the proportion of the output bucket wasted. w must fall in the range $[0, 1)$. The problem is now

$$\min_{a, b} w(a, b) \colon w(a, b) \geq 0 \quad \text{where} \quad w(a, b) = 1 - \frac{a \ln m}{b \ln n}$$

which is the same as

$$\max_{a,b} \frac{a \ln m}{b \ln n} \quad \text{such that} \quad \frac{a \ln m}{b \ln n} \leq 1$$

$$\max_{a,b} K \frac{a}{b} \quad \text{such that} \quad \frac{a}{b} \leq \frac{1}{K} \quad \text{where} \quad K = \frac{\ln m}{\ln n}$$

The result means that for a given m and n, buffer sizes a and b can be chosen so as to approximate $1/K$ as closely as desired and hence convert the streams as efficiently as desired. This is a consequence of Shannon's fundamental theorem for a noiseless channel[Shannon48]. Unfortunately as a and b increase, so does the cost of coding.

1.3.3 Impure Blocking

In the example of the previous section, the input packets all had a uniform probability of $1/m$ and so compression did not take place. For non-uniform probabilities, events of arbitrary and varying probabilities must be mapped onto uniform output buckets. Again, if $n \geq m$, input values can be mapped onto output values at high cost. Pure blocking is more efficient but does not make use of the varying probabilities.

The simplest efficient solution is to form a mapping between input strings and output strings of various lengths with the aim of matching their probabilities as closely as possible. In the input case, the probability is the estimated probability for the string. In the output case it is set at m^{-l} (where l is the length of the string) so as to maximize information content. To simplify parsing, each set of strings must possess the prefix property. This mapping technique is called **blocking**.

1.3.4 A Classification of Algorithms

The previous sections have shown that a non-trivial mapping is necessary in order to achieve efficient translation between source events (source strings) and channel events (channel strings). The technique of blocking encompasses nearly all the early data compression techniques and can be used to classify the techniques into four groups (**Figure 4**).

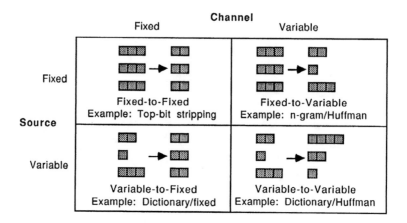

Early data compression techniques were all based upon simple mappings between source strings and channel strings. These early techniques can be divided into four groups based on the uniformity or non-uniformity of the lengths of the source and channel strings.

Figure 4: Four kinds of blocking.

Although each of these techniques can, in theory, provide optimal[19] coding, in practice variable-to-variable coding provides the most flexibility in matching the characteristics of the source with those of the channel. Later we will see how more advanced techniques enable the separation of the source and channel events that are so tightly bound in blocking techniques. For now, we continue to review the history of data compression.

1.4   Huffman Coding: Too Good Too Soon

Huffman coding marked the first major practical advance in the field. Huffman coding was born in a theoretical paper and manifested itself in a number of different practical schemes. As a result, the term "Huffman Coding" does not refer to any specific practical technique.

1.4.1   Shannon-Fano Coding and Huffman Coding

Shannon showed that for a given source and channel, coding techniques existed that would code the source with an average code length of as close to the entropy of the source as desired. Actually finding such

---

[19] Here the word "optimal" is used to mean "as close to optimal as desired".

a code was a separate problem. Given a finite set of messages with associated probabilities, the problem was to find a technique for allocating a binary code string to each message so as to minimize the average code length.

Shannon soon discovered a nearly-suitable coding technique. Fano simultaneously discovered it[20] and it has become known as the Shannon-Fano coding technique. The messages are sorted by probability and then subdivided recursively at as close to power of two boundaries as possible. The resultant binary tree when labelled with 0s and 1s describes the set of code strings.[21] This technique yields an average code length (in bits) of $[H, H + 1)$ where H is the entropy of the set of source messages.

The Shannon-Fano coding technique, though efficient, was not optimal. Very soon, Huffman proposed a variation that was optimal[**Huffman52**]. For the problem as specified, a Huffman code cannot be bettered.

To form a Huffman code, the two least probable messages are combined into a single pseudo message whose probability is the sum of the probabilities of its component messages. The pseudo message replaces the two messages in the list of messages and the grouping process is repeated iteratively until there is only one pseudo message left. The resultant binary tree describes the set of code strings.

It is a little known fact that Huffman generalized his coding technique for channel alphabets of more than two symbols. Traditionally, his code is associated with the binary alphabet, presumably because it is most easily implemented in that form.

The codes just described provide a mapping from a *set of messages* to code strings. Because it is not practical to manipulate sets of messages whose cardinality is greater than the number of atoms in the universe, Huffman coding is usually applied to *each individual instance*. Unfortunately, the minor redundancies that are a function of having to code each message into an integer number of bits are accumulated *for each instance*. This results in unbounded inefficiency if the greatest probability is close to one. This inefficiency can be reduced by using blocking to run the code at a higher level. For example, binary source

---

[20] Reportedly([Bell89]) described in R.M. Fano "The transmission of information", Technical Report 65, Research Laboratory of Electronics, MIT, Cambridge, MA, 1949. (The author of this thesis has not obtained a copy of this report).

[21] In this thesis, the two binary digits are referred to as "0" and "1". The words "zero" and "one" are reserved for expressing quantities or attributes.

streams are often organized into fixed length (say 12-bit) blocks to which Huffman coding is applied.

Unfortunately, the optimality of the Huffman code at the message level is often mistaken for optimality at the instance level.

1.4.2 Huffman as a Back-End Coder

Huffman coding has been used as a back-end coder in a variety of data compression schemes. The common theme is the division of instance streams into events with a large number of outcomes whose probabilities are roughly even. Huffman codes tend to perform well with large alphabets and this fact can be exploited by constructing source alphabets of *words* rather than characters. McCarthy[**Mccarthy73**] described a compression technique which maps strings onto Huffman codes.

Another example of Huffman as a back-end coder is described by Jakobsson[**Jakobsson78**] who divided the input stream into k-bit blocks each of which was assigned a probability based on the number of 0s and 1s that it contained. These blocks and their probabilities were then used to drive a Huffman code.

1.4.3 Multi-Group Huffman Coding

One of the more interesting variations on Huffman coding is the so-called "multi-group coding" technique. The earliest publication of the technique seems to be by Hazboun and Bassiouni in [**Hazboun82**]. Bassiouni went on to present variations of the technique in [**Bassiouni85**] and [**Bassiouni86**] (with Ok). A similar technique is described in [**Cormack85**].

The multi-group technique is designed to compress sources that generate bursts (runs) of instances from disjoint subsets of the symbol set. Such a source might generate a burst of letters followed by a burst of digits followed by a burst of spaces. Such sources are common in databases containing fields of different types.

To design a multi-group algorithm, a Huffman tree is constructed for each subset (class or group) of symbols. An extra pseudo symbol called the "failure symbol" is inserted into each tree. Each instance is coded by looking it up in the current Huffman tree. If the instance's symbol is in the tree, the symbol's code is transmitted. If the instance's symbol is not in the tree, the code for the failure symbol is transmitted to indicate that the current tree is to be changed. Separate Huffman trees are used to send a code identifying the new current tree which is then used to transmit the original instance.

A similar scheme involving multiple Huffman trees is described by Cormack[Cormack85]. In this version, all the trees contain all possible symbols and the current tree is determined solely by the most recently coded instance. This eliminates the need for failure trees. The technique was successfully applied at the device driver level in a database system and produced good compression on a wide variety of data using almost no memory.

1.4.4 Dynamic Huffman Coding

To honour the twenty-fifth anniversary of Huffman coding, Gallager [Gallager78], presented three new theorems about Huffman codes, the first of which showed that a Huffman tree with a distinguished node can be converted, in time logarithmic to the number of nodes in the tree, to another Huffman tree which would require no structural change (in order to remain a Huffman tree) were the count on the distinguished node incremented. The conversion involves swapping subtrees of equal weights. Knuth seized on Gallager's idea and detailed an algorithm which can efficiently maintain a Huffman tree whose leaf weights are being decremented as well as incremented[Knuth85]. A similar principle was used by Moffat[Moffat88] in his design of a data structure for arithmetic coding (Section 1.11.5).

The capacity to modify Huffman trees dynamically, opens up other possibilities. Moffat[Moffat87] described a two-pass technique in which the actual symbol frequencies are transmitted after the first pass. The message is transmitted on the second pass during which both sender and receiver count down the frequency of each symbol. The last instance of the message need not be explicitly transmitted.

The ability to efficiently update a Huffman tree allows Huffman codes to be altered during transmission. This means that they can be configured in the style of modern one-pass data compression techniques using the statistics of the history to construct the code for the instance or instances to follow.

1.4.5 Recent Results

The Huffman code is remarkably versatile and resilient. Despite its age, research results are still rolling in. It is worth briefly examining some of this work in order to illustrate the dominance of the Huffman code.

Johnsen[**Johnsen80**] proved that given a two symbol channel alphabet, the most probable symbol can immediately be assigned a single bit code if its probability equals or exceeds 0.4. Golomb[**Golomb80**] investigated the kind of sources that maximize the number of choices of optimal but coding-distinct Huffman trees.

Even today, interesting practical work is still being performed. In 1985 McIntyre and Pechura[**McIntyre85**] performed experiments which indicated that for small files (and in many cases large ones) two pass (so-called dynamic) Huffman coding is less efficient than (so-called static) one-pass Huffman coding which uses a fixed coding tree for all files.

In one sense, Huffman's code was too good too soon. Its proof of optimality at the message level led many to believe that it was impossible to do better, while its inefficiency at the instance level sidetracked the field into blocking. Today the Huffman optimality theorem seems rather restricted. With hindsight we can identify the two mistaken assumptions that were made on the switch from the message level to the instance level. The first was that instances in a message are independent of one another. The second was that each source symbol must be mapped onto a discrete number of channel symbols. It would be thirty years before these assumptions would be exposed and discarded.

1.5 Thirty Years of Ad Hoc Data Compression

Following Huffman's publication and the proliferation of computers, the field of data compression expanded. Although some of the work was rigorous and theoretically sound, much of it was not. It may seem a little harsh to classify *all* of the techniques to be described in this section as *ad hoc*, but in the light of the modern framework, it is hard to view them as otherwise. At least one other author agrees roughly with this categorization[**Horspool88**](end of first paragraph).

In the spirit of Huffman, this period was characterised by the simplicity of its source models and the directness of its coding schemes. Whereas contemporary compression systems code a particular message instance in a storm of probabilities, coding schemes of the 1950s, 60s and 70s are characterized by their intelligible mapping between source symbols and channel symbols. Certainly the notion of representation is more explicit in these older schemes.

Much of the research of this period focussed on the derivation of higher order source alphabets, in which each symbol corresponds to a string of ordinary source instances. Such **alphabet extensions** were used to project sources with highly skewed distributions onto flatter sources for which simple coding schemes (such as the Huffman or the fixed length codes) could more efficiently be driven. This section examines some of the ad hoc techniques that arose during this thirty year period ($\approx$ 1950–$\approx$ 1980).

1.5.1 Run Length Coding

Run length coding[22] is a technique that parses the message into consecutive sequences (runs) of identical instances. As with Huffman coding, run length coding takes many forms and has been used as a component of many compression algorithms. Run length coding can usually be identified by its trade mark of coding a run of identical data values by a single instance of the repeated value followed by a repetition count.

1.5.1.1 Binary Run Length Coding

Binary run-length codes are of particular interest because they can represent their data as a sequence of lengths of alternate 0 and 1 runs; the symbol repeated in each run need not be sent. Furthermore, because binary codes have such a small alphabet, it is likely that long runs of instances of the more frequent symbol will occur.

One of the earliest and most influential descriptions of run length coding can be found in a correspondence by Golomb[**Golomb66**]. This letter addresses the case of a binary memoryless source[23] that emits 1s with probability p and 0s with probability q (where $q = (1 - p)$ and $p \gg q$). Golomb's technique is to parse the message into runs of zero or more 1s terminated by a 0. This yields a sequence of run lengths

---

[22] It is not clear whether this form of coding should be termed "Run Length Encoding" or "Run Length Coding". The former is the name used in much of the literature. The later conforms with the names of other forms of coding (e.g. Huffman coding, arithmetic coding, adaptive coding). The later form will be used in this thesis.

[23] A memoryless source is a source that generates each instance with a fixed probability distribution and independently of all the other instances it has generated or will generate.

(a sequence of natural numbers) which are coded into binary words of varying length.

Golomb observed that the probability of a run of length $n + m$ is half that of a run of length n for $m = -\log_2 p$. From this he concluded that if the infinite set of lengths were to be mapped onto binary strings (codewords) of varying length (satisfying the prefix property) then the set of code words should contain m codewords of each possible codeword length.

Golomb proposed a code that satisfies this condition for a given m. The length L to be encoded is expressed in the form $(Q2^m) + R$. The codeword is Q expressed in base one[24] followed by R (the remainder) expressed in binary as m bits. This code satisfies the required property because the number of bits required to transmit the remainder R is fixed for a given m and the number of bits needed to transmit the quotient grows by one bit for each increase in run length of m. A slightly more complicated version caters for values of m that are not powers of two. This Golomb code forms a subset of a particular arithmetic code developed by Langdon and Rissanen[**Langdon82**].

Despite its age, the technique of run length coding is still being studied. In a comprehensive paper[**Tanaka82**], Tanaka and Leon-Garcia described a form of run length coding that is highly efficient for binary memoryless sources. They defined a mode m code to be a code that maps a run of 0s of length m into one bit (0) and runs of length $0 \ldots 2^m - 1$ into $m + 1$ bits (1 . . .). The input is divided up into blocks of L bits. Each block is processed separately. The number of 1s in the block is counted and the probability of a 0 is estimated. The optimal m for this p is calculated and the block is encoded with a mode m code. The authors prove that the largest possible difference between the coding rate and the corresponding entropy is 0.0405 bits. Typically, the efficiency is 99% of the entropy. Coding is fast because the order m codes can be precomputed.

A remarkable similarity exists between the efficiency curves of Tanaka's and Leon-Garcia's code (figure 1 of [Tanaka82]) and that of Langdon and Rissanen's binary arithmetic code (figure 2 of [**Langdon81**]). Both curves consist of a series of humps of exponentially

---

[24] Base one uses only one digit (1). Numbers are represented by the *number of occurrences of the digit*. For example the decimal number 3 is represented by 111 in base one. In the case of the Golomb code, the base one representation is terminated by a 0 so that the end of the number can be detected in the binary code stream. Thus 1 is represented by 10 and 3 by 1110.

decreasing width which represent the "modes" at which the code can operate; both have a lower bound in efficiency of about 96%. Langdon suspects that the similarity is because both techniques assign an integer-length code string increase to the less popular symbol.[25] This similarity shows well how, in the field of data compression, information properties can be harnessed in radically different ways to the same effect. We will see more of this in the section on Markov algorithms (Section 1.10).

Bahl and Kobayashi[**Bahl74**] presented a scheme for coding a binary memoryless source for an image coding application. In particular, two of the schemes cover two of the four blocking classes (Section 1.3.4) and are worthy of further description.

The first coding scheme is a variable-to-fixed run-length code. The probability p of the most likely digit is used to select an N. Then, N-bit words are transmitted which contain the lengths of successive runs of 0s and 1s. If the run length exceeds the capacity of the word size, the maximum value is transmitted and the counter is reset. An alternative method is to transmit a sequence of the lengths of "runs" defined to be sequences of zero or more 0s followed by a single 1.

The second scheme is a more complicated variable-to-variable run-length code. It suffices to say that the run lengths are coded using varying-length binary codes with the shorter run lengths assigned shorter codes than the longer run lengths. A further refinement, called multi-mode Golomb coding, assigns shorter codes to the *more probable* run lengths. Because a memoryless source produces a geometric run-length distribution the shortest codes are assigned to runs with near average length rather than to runs with shorter lengths.

Teuhola[**Teuhola78**] described another modification of the Golomb code in which runs of instances of the more probable symbol are parsed into subruns of length $2^k, 2^{k+1}$ for some predefined base power k determined by p. The number of such complete subruns is transmitted in base one and the length of the leftover is transmitted in j bits where $\log_2 j$ is the length of the last complete subrun. This scheme is fairly insensitive to the base power k because of the "binary exponential backoff". It is also simpler than multi-mode Golomb coding.

---

[25] Private correspondence in a mail message 12 August 1988. Permission to quote was granted in another mail message on 6 May 1989.

1.5.1.2 Bit Vector Compression

A number of other techniques for compressing binary memoryless sources are worth mentioning even though they do not use run-lengths explicitly. Such techniques are sometimes referred to as **bit vector compression techniques**. Whereas run-length coding relies directly on the occurrence of runs of symbols, bit vector techniques rely on the increased occurrence of the more frequent symbol.

A technique described by Jakobsson[Jakobsson78] divides the input bit stream into k-bit blocks each of which can be assigned a probability based upon the number of 0s and 1s. These blocks and their probabilities are used to drive a Huffman code. Jakobsson analyses the scheme and shows that good compression can be achieved even for k as low as 10.

In a later paper, Jakobsson[**Jakobsson82**] described a similar blocking technique. The source bit stream is parsed into blocks of k bits and an index is constructed with one bit corresponding to each block. Each bit of the index is set to 0 if its corresponding block is all 0s. The source is then coded by sending the index followed by the non-zero blocks. Before this takes place, the index is coded in the same way. This process repeats iteratively until a predetermined level count is reached.

1.5.2 Dictionary Techniques

One of the most obvious redundancies of many data sets (and in particular text files) is the repeated occurrence of substrings. For example, a particular identifier may be referred to in a program text many times. It is therefore not surprising to find that a great many data compression techniques have been based on the detection and elimination of these repeated strings.[26] Techniques that "factorize" common substrings are known as "dictionary techniques".

Dictionary techniques construct a dictionary of common substrings either on the fly or in a separate pass. Channel instance strings are associated with each dictionary entry and the message is transmitted by parsing the message into dictionary entries and transmitting the corresponding channel strings. Dictionary techniques can be fixed-to-fixed (e.g. entries are all the same length and are mapped to fixed length codes), fixed-to-varying (e.g. dictionary consists of n-grams,[27] which are coded using a Huffman code), varying-to-fixed (e.g. dictionary contains

---

[26] The term "string" is usually taken to mean a text string. In this context the term includes arbitrary byte streams. By "string" is meant "sequence of source instances".
[27] An n-gram is any sequence of instances of length n.

different length words which are coded into 12 bit integers) or varying-to-varying (e.g. dictionary contains arbitrary strings which are coded using Huffman coding) (**Figure 4**).

1.5.2.1 Parsing Strategies

Given a dictionary and a message, there are many ways in which the message can be expressed in terms of the dictionary. For example, if the dictionary contained the strings listed in **Table 2**, and the message was wooloomooloo, the message could be **parsed** in any of the following ways.

```
woo-loo-moo-loo
woo-loom-ooloo
woo-loomooloo
wool-oom-ooloo
wool-oomoo-loo
```

Dictionary
loo
loom
loomooloo
moo
ooloo
oom
oomoo
woo
wool

This dictionary, which is used in the wooloomooloo example has been specially designed to highlight the ambiguity of the parsing problem. This dictionary does *not* have the prefix property (Section 1.2).

Table 2: Example dictionary.

The way in which a message is parsed affects compression. In the example, if each word was mapped to a fixed length code, a two-word parsing would use half the space of a four word parsing. A number of different parsing algorithms exist which vary in speed and efficiency.

An optimal algorithm can be constructed by mapping the parsing problem onto a shortest path problem and solving that (**Figure 5**).

The arcs of the graph are labelled with dictionary entries (with their associated codeword length cost). The nodes correspond to positions in the message. The approach is more efficient than it might sound because for a given dictionary, most strings have **cut points** at which a parsing division must occur. In the example, a cut point would occur at woo-loomooloo if the word wool were not in the dictionary. The end points of the message are always cut points.

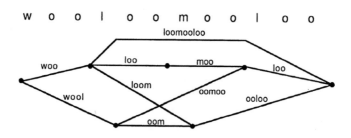

For a given message and a given dictionary, the parsing problem can be mapped onto the shortest path problem by constructing a directed acyclic graph of all possible parses. The graph can then be solved in linear time using dynamic programming. This graph corresponds to the wooloomooloo example.

Figure 5: Parsing problem mapped onto shortest path problem.

The problem of optimal parsing is similar to the problem faced by the TeX typesetting system of breaking paragraphs into lines[Knuth79] [Knuth84]. TeX uses dynamic programming to solve the problem. Dynamic programming was earlier proposed as a solution to the parsing problem by Wagner[**Wagner73**] who showed that for a given dictionary, optimal parsing could be performed in time linear in the length of the text to be parsed. It should be noted that the problem of choosing an optimal *dictionary* for a given text is NP-complete[**Storer82**].

Although optimal parsing can be implemented with reasonable efficiency, much faster techniques exist that perform nearly as well. The LFF (longest fragment first) heuristic examines the entire string and iteratively replaces the longest matching substring with its codeword. By far and away the most popular parsing algorithm, however, is the "greedy algorithm", which works sequentially through the string parsing at each step the longest dictionary entry that matches the next few instances in the message. In the example above, the greedy parsing algorithm would parse wooloomooloo as wool-oomoo-loo. Greedy parsing is not optimal; by parsing the longest phrase first (wool), the algorithm misses out on the big prize (loomooloo). In practice, greedy parsing performs nearly as well as optimal parsing but is simpler and faster[**Schuegraf74**]. Unless otherwise stated, all the dictionary techniques reviewed in this chapter use greedy parsing.

1.5.2.2 Static vs Semi-Adaptive Dictionaries

In their book on text compression[Bell89], Bell, Cleary and Witten divided dictionary techniques into static and semi-adaptive techniques. Static techniques use the same dictionary for all files. This enables the dictionary to be embedded in the compression program; it need not be transmitted. Semi-adaptive techniques construct a different dictionary for each message and transmit the dictionary along with the coded message. Static techniques are one pass and semi-adaptive techniques are two pass.

Unfortunately, much of the literature on dictionary techniques does not distinguish between static techniques and semi-adaptive techniques with many authors failing to specify clearly whether the technique they are describing transmits a dictionary.

1.5.2.3 Early Dictionary Techniques

The earliest work on dictionary techniques seems to be by Schwartz[**Schwartz63**] who described a greedy-parsing static-dictionary technique that replaces words (in a text) matching dictionary entries with fixed length codes; text that doesn't match is transmitted verbatim. The theoretical basis of the technique is that 500 to 1000 well chosen words will cover about 75% of most English texts. This early work is interesting because it employed a "split dictionary" in which one part contains word roots and the other part contains word endings. This organization resulted in a small dictionary that could synthesize a large number of long words. The authors discuss techniques for automatically improving the dictionary given extra training texts.

Four years later, White[**White67**] presented a similar, greedy-parsing static-dictionary technique that used fixed length codes. The dictionary consisted mostly of highly probable English words but also contained special strings, capitalized words and suffixes. The technique contained a lot of ad-hoc tuning to cater for the specialized class of texts being compressed. The authors concluded that a dictionary containing 1000 words will yield about 50% compression for English text.

Other early work on dictionary coding was done by Notley[**Notley70**] who described a multi-pass algorithm to construct a dictionary of commonly occurring substrings (a "Cumulative Occurrence Library"). The technique was designed for authorship fingerprinting but Notley also admits to a data compression application. On each pass, greedy parsing is used to find the longest matching dictionary entry. If there is no matching entry, a new symbol is added to the dictionary. If there is an entry, the entry's count is incremented. If a count reaches a threshold, a new entry consisting of the concatenation of the entry and the previously parsed entry is inserted into the dictionary. Entries with a low count are discarded. The dictionary develops during the sequence of passes. The message is coded by transmitting the dictionary and then the message coded using fixed length coding. Mayne and James[**Mayne75**] developed this technique further by investigating heuristics for adding and removing dictionary entries.

Many of the earlier dictionary techniques for text compression use words as dictionary entities. Clare, Cook and Lynch[**Clare72**] noted that the words of natural language texts exhibit a Poisson rank-frequency distribution (see also [Zipf49]) with a small number of common words accounting for a large proportion of word occurrences. If a fixed length coding scheme is used, dictionaries tend to become less efficient the bigger they get. The solution proposed was to use arbitrary text fragments as dictionary entries rather than words.

One of the strangest dictionary techniques employs both a dictionary and run length coding. Lynch[**Lynch73**] achieved good compression using a two pass technique. In the first pass, each instance of the message is replaced by a fixed-length code, whose 0 bit content increases with the probability of the instance's symbol. The most frequent symbol is represented by 00000000 and the least frequent symbol by 11111111. The second pass performs run length coding on the result. This technique was extended to use digrams and 12-bit codes.

Other early work was done by McCarthy[McCarthy73] who described a technique for constructing a dictionary from a sample of the file to be compressed. McCarthy went to more effort to make sure that the dictionary and the consequent encoding were closer to optimal. The dictionary construction procedure iteratively selects strings according to the compression they will yield. Optimal parsing is implemented using the dynamic programming technique described in [Wagner73]. The resultant parse is used to drive a Huffman code.

Rubin[**Rubin76**] continued McCarthy's approach, experimenting with different methods for building the dictionary. Rubin constructed the dictionary iteratively, parsing the message on each pass and maintaining occurrence statistics. After each pass, the dictionary was refined by "augmenting" (adding symbols to) the "best" (according to some metric) entries and inserting them to the dictionary. Rubin tried a number of metrics for "best" based on length and frequency and concluded that frequency alone was the best measure. Three augmentation schemes were implemented: prepend a symbol, append a symbol and append an entry. The last method worked best, presumably because adding a symbol to the start or end of a group can simply rob the symbol from an end of another group (in the message string). Rubin tried coding the dictionary entries using Huffman coding but found it ineffective because the dictionary usually contained roughly equiprobable entries.

1.5.2.4 Later Dictionary Techniques

Wolff[**Wolff78**] described a static dictionary technique in which the dictionary is constructed during multiple passes over the text. Each pass adds a new entry to the dictionary. The number of passes depends on how large a dictionary is required. Before the first pass, the dictionary is primed with the character set. On each pass, greedy parsing is used to parse the message string into dictionary phrases. During the pass, a count is kept of the frequency of all pairs of phrases. At the end of the pass, the most frequently occurring pair are added to the dictionary. The message is coded using a binary fixed length code.

Cooper and Lynch[**Cooper78**] described a greedy-parsing/fixed length code dictionary technique for compressing files containing chemical structure information (Wiswesser Line Notation).

Weiss and Vernor[**Weiss78**] described a dictionary technique that employs a dictionary of 1024 text words. The technique replaces words in the message text by sixteen-bit codes. Text that does not match a dictionary entry is included verbatim. Two of the spare six bits in the dictionary pointers are used to indicate the presence of a pointer, one is used to indicate that the word's first letter is capitalized and three are used to describe how the coded word ends (e.g. question mark, comma).

Because dictionary compression techniques require a lot of table searching, an investigation was made into how hardware could be employed to perform this task more quickly. Lea[**Lea78**] described an associative memory that could be used to eliminate the table searches in dictionary compression schemes.

Compression can be achieved by finding the m most frequently occurring n-grams (typically n is 2 or 3). Yannakoudakis, Goyal and

Huggill[**Yannakoudakis82**] described such a method. They note that this method is "an attempt to convert the normal hyperbolic distribution of single letters to a rectangular or equiprobable distribution of symbol groups (n-grams) by considering frequently occurring strings of up to n letters in length." (p. 17) Again the assumption of fixed length channel strings is made.

Surprisingly good compression has been achieved by applying Huffman coding to the output of a dictionary compression scheme[Bassiouni86]. This is an unusual approach because compression algorithms aim to produce white noise. In this algorithm, the second pass catches some redundancy missed in the first. The first phase consists of a dictionary compression scheme that augments the dictionary entry numbers with bit fields to indicate capitalization and various common endings. Run length encoding is used as an alternative to referring to a dictionary entry. The second phase uses a multi-group technique (Section 1.4.3) to remove redundancy in the output of the first pass. Although particularly ad-hoc, this technique removes approximately 65% of English text files and illustrates how the combination of a number of techniques can sometimes be effective.

1.5.2.5 Compressing the Dictionary

Once the dictionary has been constructed and the message encoded, it remains only to transmit the dictionary and the coded message. A final opportunity remains for compression if the dictionary itself can be compressed. One way of doing this is to construct a string of overlapping dictionary entries and store each entry as a pointer into the string. Gallant analysed the problem of finding a minimal length superstring of a finite set of substrings[**Gallant80**] and proved that it was NP complete. Efficient algorithms may exist for this problem but it is likely that they would be neglected in practice for the same reasons that optimal parsing is neglected.

The problem of compressing the dictionary can be avoided by organising the transmitter and receiver to simultaneously build identical dictionaries as compression proceeds. This idea has been incorporated into the class of Lempel-Ziv compression algorithms. Lempel-Ziv algorithms are theoretically based, fast and have so completely surpassed the dictionary techniques just described that discussion of them is deferred to a section specifically devoted to them (Section 1.7).

1.5.3 Exploiting Specific Knowledge

The techniques described so far are fairly general. Most of them will perform effectively on a variety of data. However, if more information is known about the data to be compressed, the compressor can be tailored for the specific situation. Examples of specific data sets are computer programs, English text, bibliographic data, chemical data, signal data and sparse databases.

If detailed information is available about the structure of messages, a specific compressor can be constructed which exploits that knowledge. Katajainen, Penttonen and Teuhola[**Katajainen86**], described a method for compressing program files in which a parse tree is constructed and coded using Huffman coding. A better technique is to construct a probabilistic grammar for the set of source messages and code each message as a sequence of syntax graph decisions[**Stone86**]. Such a technique could be embedded in programs that automatically generate interactive programming environments[**Reps84**].

1.5.4 Data Base Compression Techniques

The requirements of data base representation lead to a perspective of data compression slightly different from that of mainstream data compression research. Data base compression differs from ordinary file compression in two ways. First, data bases tend to be sparse and are unusually easy to compress. Second, there is usually the need to be able to access the data quickly and at random. These characteristics tend to constrain the set of possible data compression techniques to those that provide a simple mapping. An example is the compression of file indexes that must remain accessible in their compressed state. This section describes some of the techniques used to compress data bases.

In a sequence of sorted items, it is usual to find that adjacent elements share a common head string. The technique of **differencing** reduces the length of each element by replacing the head string that each shares with the previous element by a number that is the length of the shared head string. This is called front-compression. In rear-compaction[28] as much of the tail of a word is removed without making it identical to the previous word. Rear compaction is capable of losing information and is suited only to specialist indexing applications.

---

[28] A *compression* technique never loses information. A *compaction* technique can sometimes lose information[**Severance83**][**Reghbati81**]. This thesis is concerned only with compression techniques.

Eggers and Shoshani[**Eggers80**] presented a method for compressing large, sparse data bases. In a sparse data base, one particular constant value fills most of the data base. The first stage of compression is to serialize the data base. The constant is then deleted from the data stream and a map is constructed containing the location of the deleted runs and the remaining data. This operation can be repeated to eliminate other constants in the remaining data. This technique allows search techniques with a logarithmic access time (e.g. binary search) to be used on the compressed data base. The second part of the paper describes how this can be extended to the case of databases not all of whose keys are used.

Other ad-hoc techniques can be used to great effect. The printable members of the ASCII[29] character set utilize only seven of the eight bits of the standard ASCII code and it is possible to store each character in only seven bits. If lower case characters are unused, only six bits are required. By eliminating these bits, text storage can be reduced by 25%. Fixed length text data is notorious for containing trailing blanks. Trailing blanks can be eliminated by deleting them (i.e. compaction, if the definition of a text file allows it) or by replacing them using run-length coding. Another method for reducing "whitespace" is to use tabbing, in which runs of blanks are replaced by a single tab character.

Scientific data bases offer enormous scope for compression[Bassiouni85]. They often contain long sequences of almost identical numbers (taken perhaps from a slowly varying instrument) which can be compressed using a differencing technique.

In general, the techniques used to compress data bases are oriented around recovery speed and the properties of the specific data being compressed. Data bases are usually so easy to compress that the extra effort required to shave off an extra few percent is usually not considered worthwhile. As a result, techniques developed to represent data bases are not at the frontier of the field.

1.5.5   The Practitioner's View

With so many compression techniques available, it is interesting to observe the reaction of practitioners in the field whose job it is to manage large bodies of data.

There is an abundance of literature which reviews data compression techniques (e.g. [Reghbati81], [**Gottlieb75**], [**Cooper82**], [Severance83] [**Peterson79**]). Most of these works describe a subset of the ad-hoc

---

[29] American Standard Code for Information Interchange.

data compression techniques and follow it up with a brief comparison in which a particular point of view is put forward. Most papers do not even mention the modern techniques described in Section 1.8. For example, a recent *review* paper[**Lelewer87**] in a *review* journal described static and dynamic Huffman compression in great detail, briefly described LZ78 and Bentley, Sleator, Tarjan and Wei's move-to-front scheme[**Bentley86**] but only *mentioned* Markov modelling (a major focus of this thesis) in the fourth last paragraph of the paper. The issue of modelling was hardly addressed, as Witten and Bell pointed out in their review of the paper[**Witten89**]. Hopefully the modern paradigm will become well known in a few years. Some up-to-date works are appearing[Helman88] [**Witten87**][**Abrahamson89**].

With so many techniques to choose from and few formal guidelines, it seems that the reaction of the average data practitioner has been to ignore data compression altogether! Welch[**Welch84**] gives three reasons for the neglect of data compression: poor execution speed, the inability of most techniques to compress different kinds of data, and the unpredictability of the size of compressed data. Severance [Severance83] also gives three reasons: designers underestimating the compression possible, wariness of the extra complexity introduced by a data compression layer, and the narrowness and mathematical mystique surrounding the data compression literature.

> "What is clear nevertheless is that typical commercial databases can in fact be compacted by 30–90%, and that this should be of more applied interest than current usage of compression techniques would indicate." [Severance83]

It is of significance that most reviewers complain that data compression techniques are not used as much as they should be. Perhaps the point to be made is that *some* data compression is better than none. Indeed, as the greatest gains are to be made by taking the simplest measures, it is surprising that data compression techniques are not already in wide practical use. As simple a technique as null suppression can reduce a data base by up to 70%[Severance83]. In addition, there are indications that data compression does not significantly impact on processing efficiency and in some cases can actually improve it[**Smith76**]. This is in contrast to the commonly held belief that data compression will slow one's system down.

Another problem with data compression is deciding how it is to be used. Data compression mechanisms can be located in disk drivers, in ordinary programs (e.g. editors), in separate utilities that must be explicitly invoked or in background jobs. Choosing the location can be difficult. Perhaps one of the most aggressive implementations is described by Raita[**Raita87**] whose program hunts down and compresses users' files that have not recently been accessed.

To summarize, there is a very large gap between data compression research and practice. This gap can only be closed by commercial pressures (as has happened with facsimile compression) or by researchers presenting a simpler, more concrete image of data compression.

1.6   Adaptive Data Compression

All the techniques so far described either use fixed statistics, or make two passes over the message, the first pass to gather statistics and the second pass to code the message (using the statistics). Bell, Cleary and Witten[Bell89] called the fixed statistics techniques "static" and the two pass techniques "semi-adaptive".

Static and semi-adaptive techniques are unsatisfactory for general-purpose data compression. Static techniques cannot adapt to unexpected data, and semi-adaptive techniques require two passes, making them unsuitable for communication lines.

"Adaptive" techniques combine the best of static and semi-static techniques by making a single sequential pass over the message, adapting as they go. At each step, the next piece of message is transmitted using a code constructed from the history. This is possible because both the transmitter and receiver have access to the history and can independently construct the code used to transmit the next piece of the message.

An example of an adaptive technique is a technique in which each instance is transmitted using a Huffman code constructed from the history. A straightforward implementation would be prohibitively inefficient because it would require that a new Huffman tree be constructed from the history for each instance transmitted. The trick is to design a data structure that can be incrementally updated at low cost. Such a dynamic technique has already been devised for Huffman coding (Section 1.4.4).

The advantages of adapting during a single pass might be considered reason enough to dispense with static and semi-adaptive techniques altogether. In fact, arguments exist that permanently lay the question to rest.

A static technique will always yield better compression than an adaptive technique on the data for which the static technique has been tuned. This is because the adaptive technique must spend time learning what the static technique already "knows". The best that could be expected under these conditions is that adaptive algorithms use up at most M nats more channel space for any message than the best static model, where M is the amount of information contained in the static model. Such a theorem has been proven under fairly general conditions [Cleary84] (in particular, the source must be ergodic). Static models exhibit an unbounded inefficiency if they are fed unsuitable data. Thus, the only advantage that static models have over dynamic models is a constant-order compression advantage for the particular kind of data for which they were tuned.

The same theorem shows that adaptive techniques are superior to semi-adaptive techniques. At best, a semi-adaptive technique can construct and transmit the optimal static model for the given data. Transmission of the model will cost M nats. If an adaptive model uses no more than M nats more than the best static model (the proof of which was discussed in the last paragraph) then it can perform no worse than the semi-adaptive model. Thus semi-adaptive models have *no* advantage over adaptive models.

The conclusions above enable us to add a new assumption to our view of data compression.[30]

Assumption 5: Compression takes place in a single pass; the transmitter can only see a small finite part of the remainder of the message.

1.7 Ziv and Lempel Algorithms

1.7.1 Adaptive Dictionary Compression

All dictionary techniques have to find some method of transmitting the dictionary from the transmitter to the receiver. The dictionary techniques that we have seen so far do this either by using a static dictionary which doesn't need to be transmitted (static techniques) or by transmitting the dictionary before transmitting the message (semi-adaptive techniques).

Adaptive dictionary techniques do not ever transmit the dictionary explicitly. Instead the transmitter and receiver both build the dictionary incrementally, adding to it as each instance (or group of instances) is transmitted. At each point during the coding, the current dictionary is used to transmit the next portion of the message.

---

[30] The previous assumptions are listed in Section 1.1.3.

The notion of sender and receiver simultaneously maintaining identical models of the source is fundamental to the modern paradigm of data compression (Section 1.8). Thus, even though Ziv and Lempel techniques are dictionary based, it would be unfair to classify them as ad-hoc techniques.

1.7.2 The Launch of LZ Compression

Ziv and Lempel coding (LZ coding[31]) refers to two distinct but related coding techniques first presented by Ziv and Lempel in two papers published in 1977[**Ziv77**] and 1978[**Ziv78**]. The fundamental idea behind LZ algorithms is that substrings of the message are replaced by a reference (e.g. an (offset,length) tuple) to a substring in an earlier part of the message.

Ziv and Lempel's approach was to hide a good idea in a sea of mathematics. Bell, Cleary and Witten restate this feeling in gentler terms, summarizing the problem well:[32]

> "It is a common misconception that LZ coding is a single, well-defined algorithm. The original LZ papers were highly theoretical, and subsequent accounts by other authors give more accessible descriptions. Because these subsequent descriptions are innovative to some extent, a very blurred picture has emerged of what LZ coding really is. With so many variations on the theme, LZ coding is best described as a growing family of algorithms, with each member reflecting different design decisions." [Bell89] (section 8.3(v))

In order to avoid such confusion, the two original algorithms will be described in detail, followed by a discussion of their variations. A more detailed discussion and explanation of LZ algorithms and all their variations can be found in the book by Bell, Cleary and Witten[Bell89] which introduces a naming scheme for the algorithms based on the name of the authors. The two algorithms by Ziv and Lempel are named LZ77 and LZ78. Mathematical notation used in the following descriptions does not follow the original papers, but rather the book whose notation is simpler.

---

[31] According to [Bell89], the technique is usually referred to as "Ziv and Lempel coding" (after the ordering of the authors' names in the original papers) but when acronymized is referred to as "LZ" coding (because of a historical mistake).

[32] Quotes from and comments on the book [Bell89] refer to a near-final draft of the book printed on 16 May 1988. Permission to quote from the draft (conditional on stating that it was a draft) was granted by Tim Bell in an electronic mail message to the author of this thesis on 19 December 1988 for which appreciation is recorded.

1.7.3 LZ77

The LZ77 algorithm takes two parameters:

- $N \in \mathbf{Z}[1, \infty)$, the length of a sliding window buffer.

- $F \in \mathbf{Z}[1, N - 1]$, the maximum length of a matching string.

with $F \ll N$. Typical values that are used in practice are $N \approx 2^{13}$ and $F \approx 2^4$; as usual, powers of two encourage an efficient implementation.

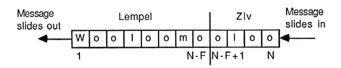

The LZ77 algorithm slides the message through a buffer from right to left. The Lempel holds the text already transmitted and the Ziv holds the text to be transmitted. At each step, the next part of the Ziv is transmitted to the receiver by expressing it as a substring of the Lempel using an offset and a length.

Figure 6: The LZ77 algorithm.

Execution of LZ77 revolves around a sliding window buffer of length N through which the message is passed from right to left. The buffer's elements are numbered consecutively with 1 at the far left and N at the far right. We will call the $N - F$ leftmost elements (elements $[1, N - F]$) the **Lempel** and the F rightmost elements (elements $[N - F + 1, N]$) the **Ziv**.[33] The Lempel holds the most recent $N - F$ instances that have been transmitted and the Ziv holds the next F instances to be transmitted.

To start, the algorithm initializes the Lempel to a pre-defined string and slides the first part of the message string into the Ziv. Coding proceeds by finding the longest substring in the buffer whose leftmost element lies in the Lempel and which matches the first zero or more instances in the Ziv, and transmitting it (along with the next instance a) as the triple (s, l, a) where $s \in \mathbf{Z}[1, N - F]$ is the position in the Lempel where the matching string starts,[34] $l \in \mathbf{Z}[0, F]$ is the length of the matching string and $a \in A$ is the instance following the matching string. The message is then slid into the buffer from the right until

---

[33] Warning: This *Ziv* and *Lempel* terminology is unique to this thesis.

[34] This is traditionally indexed from the rightmost position $(N - F)$ of the Lempel with $(N - F) \rightarrow 1$.

the next instance to be encoded is at the leftmost element of the Ziv (i.e. element $N - F + 1$).

The following notes reveal how ingenious the algorithm really is:

• The extra instance a is transmitted to cover the case that no match is found (i.e. when $l = 0$).

• The matching string can start near the end of the Lempel and extend into the Ziv. This works because the decoder will have reconstructed the part of the matching string in the Ziv by the time that section itself must be copied. This feature means that the algorithm efficiently codes runs of identical instances.

• The "dictionary" for this technique consists of every substring in the Lempel. Despite this, it is never explicitly transmitted because it is updated incrementally by both the coder and the decoder.

• The algorithm is locally adaptive (Chapter 3) because its model is based solely upon the previous $N - F$ instances.

• Searching the buffer for the longest matching string is expensive but bounded by N and F. The algorithm codes and decodes in time linear in the length of the message.

• Decoding is extremely fast. The decoder uses a buffer identical to the coder and repeatedly copies the substrings specified by the stream of triples from the Lempel to the Ziv. The message comes out the leftmost side of the Lempel.

• Because N and F are finite, s, l and a can all be packed into fixed-length bit fields.

• The sliding window can be implemented using **mod** N arithmetic (Section 4.5), which eliminates the need for explicit buffer sliding which becomes expensive for large N.

• Ziv and Lempel showed that LZ77 could perform at least as well as a semi-adaptive dictionary technique.

1.7.4 LZ77 Variants

Bell, Cleary and Witten[Bell89] identified four LZ77 variants. Rather than enumerate them, we will briefly mention the modifications they introduced.

Infinite Window Width: LZ77 uses a fixed length window which means that its model is always based on the previous $N - F$ instances seen. However, if N is set to ∞, the model is based on the entire history. For such a high N, transmitting s (the offset) becomes a problem which can be solved by using a varying length code for s in which an integer of value i is coded in $O(\log_2 i)$ bits.[35] This means that the more recently a string has occurred, the shorter is its code string. With the window width limit removed, one might suspect that a straightforward implementation of the technique would have a quadratic time complexity in the size of the input message. However, Rodeh, Pratt and Even[**Rodeh81**] showed that sophisticated data structures can reduce the time cost to linear complexity.

Infinite Maximum Matching Length: The maximum matching length F can be similarly generalized.

Eliminate Instance: Transmitting an instance (a) as the third element of a triple is wasteful if the instance could have appeared as part of the next triple. In one variant, each tuple is preceded by a bit which specifies whether (a) or (s,l) is to be transmitted. At each step, the coder chooses the alternative that will most compactly represent the matching substring.

Variable width pointers: Brent[**Brent87**] described a technique in which the results of a variant of LZ77 are coded using a Huffman code.

1.7.5 LZ78

The LZ78 algorithm is similar to LZ77 except that the Lempel is replaced by a continually growing dictionary of $d \in Z[1, \infty)$ **phrases** (strings) numbered from 0 to $d - 1$. No limit is placed on the length of the Ziv. The algorithm has no parameters.

---

[35] A good discussion of this form of integer coding can be found in Appendix A of [Bell89].

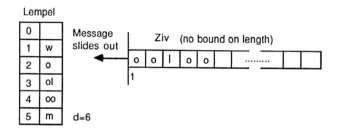

The LZ78 algorithm builds a dictionary of *phrases* and repeatedly transmits the number of the longest dictionary entry matching the Ziv, followed by one instance after that. After each phrase is transmitted, a new phrase is added to the dictionary. The new phrase is the phrase transmitted plus the following instance. This diagram depicts the state of the algorithm midway through the processing of the string wooloomooloo.

Figure 7: The LZ78 algorithm.

To start, the algorithm initializes the dictionary to a single phrase consisting of the empty string and sets $d \leftarrow 1$. At each step, the algorithm transmits a new phrase $p \in S$ in the Ziv consisting of the longest matching phrase $m \in S$ in the dictionary plus the next instance $a \in A$. Thus $p = ma$ and $Ziv = ``p \ldots"$. p is transmitted as m's dictionary entry number (requiring $\lceil \log_2 d \rceil$ bits) followed by a which is transmitted raw. The new phrase p is then inserted into the dictionary, some more of the message is slid into the Ziv, and the process repeats.

The effect of this is to parse the input into phrases, each of which consists of the longest previous phrase plus one instance. Ziv and Lempel proved that this technique converges on the entropy of a stationary ergodic source as the message length tends to infinity. While this result is important theoretically, convergence is so slow that the property is meaningless in practice. Bell, Cleary and Witten[Bell89] calculated that for a symbol set of 256 the technique will still be 20% inefficient when $d = 2^{40}$.

Here are some important features of the algorithm:

• The dictionary in this algorithm can be efficiently implemented using a digital search tree (**Figure 8**). Each parsing step involves travelling from the root to the node corresponding to m, and then attaching a new node (corresponding to p) to m with an arc labelled a.

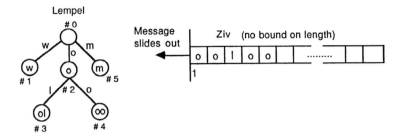

The dictionary (Lempel) of the LZ78 algorithm can be organized as a forwards tree. At each step, the algorithm transmits the number of the tip of the longest matching branch. The new phrase is then added by appending a new leaf to the tip node. This diagram depicts the state of the algorithm midway through the processing of the string wooloomooloo.

Figure 8: The LZ78 algorithm implemented using a tree.

• As it stands, the size of the dictionary increases forever. In practice, memory will eventually run out and some action must be taken. A common solution is to empty the dictionary and continue.

Langdon[**Langdon83**] cast light on the underlying mechanism of LZ78 by noting that following its creation, each node in the tree gains one descendent[36] for each occurrence of the node's phrase (or an extension thereof) in the subsequent sequence of phrases. Because the code space is divided evenly between nodes, LZ78 is really using a statistical technique at the phrase level.

Langdon showed that the actual probability that could be assigned to each arc of the LZ78 tree is c/p where c and p are the number of descendents of the child and parent nodes of the arc. In the wooloomooloo example (**Figure 8**), the arcs from o to ol and from o to oo have probability 1/3, as does the imaginary arc from o to o (i.e. stop at that node).

1.7.6 LZ78 Variants

Bell, Cleary and Witten[Bell89] identified six LZ78 variants. Rather than enumerate them, we will briefly mention the modifications they introduced.

---

[36] Here, the set of "descendents" of a node r is identical to the set of nodes contained in the subtree rooted in r. The set includes r.

Eliminate Instance: The explicit transmission of the last instance a of each new phrase p can be avoided by priming the dictionary with the symbol set and transmitting a as the first instance of the next phrase [Welch84]. This means that the decoder has to decode the $k + 1$'th phrase in order to insert the k'th phrase into the tree. A special case arises when the $k + 1$'th phrase's m is the same as the k'th phrase, which can be handled by noting that the k'th phrase's a must equal its m's first instance. Eliminating the transmission of a was the most important modification made to LZ78. A highly optimized version of this algorithm forms the core of the popular *compress* program that runs under the UNIX[**Ritchie78**] operating system. An innovative feature of *compress* is that when it runs out of memory, it freezes the dictionary and empties it only when compression performance drops.

Dictionary Management: When the dictionary becomes full, some variations (e.g. [**Tischer87**]) remove the Least Recently Used (LRU) phrase to make room for the new phrase, thus introducing local adaptivity similar to that of LZ77.

Rapid Phrase Growth: The LZ78 algorithm grows its dictionary phrases one instance at a time. In an effort to increases the rate of phrase growth, one technique inserts the concatenation of the $k - 1$'th and k'th phrase into the dictionary at each step rather than inserting just the k'th phrase. Another technique inserts not only the new phrase into the tree, but all substrings less that a maximum length whose last instance falls within the phrase in the message. When the dictionary is full, phrases that have only been used once are removed and compression continues. Eventually the dictionary becomes static.

Window/tree approach: The highest performance LZ algorithm to date is the LZFG[**Fiala89**] algorithm which has aspects of both LZ77 and LZ78. The message is passed through a sliding window buffer as in LZ77, but the instances in the window are parsed into phrases as in LZ78. A tree is maintained that contains only those phrases in the window. Coding takes place by transmitting (offset,length) pairs (as in LZ77) that match the buffer starting at a phrase boundary. This variation is fast, gives good compression and manages memory effectively.

More detailed discussion of and references to LZ77 and LZ78 and their variants can be found in the book by Bell, Cleary and Witten[Bell89] from which much of the information in this section was obtained.

Ziv and Lempel have continued and extended their work in data compression. In [**Lempel86**] they follow up that work by extending their results to two dimensional data (e.g. bitmaps).

1.7.7 Other Dynamic Dictionary Techniques

A few adaptive dictionary techniques have arisen that don't quite fit into the Ziv and Lempel mould. Bentley, Sleator, Tarjan and Wei[Bentley86] described a one-pass locally adaptive technique that maintains a list of words. No strict definition of a word is given. At each step, the next input word is compared with all the words in the list. If the next input word is in the list, the number of the word in the list is transmitted and the word is moved to the head of the list. If it isn't in the list, the word is transmitted explicitly, preceded by a number that is one more than the number of elements in the list, and then the word is added to the list. The scheme requires that the word numbers be transmitted in space proportional to the logarithm of the number. Ryabko[**Ryabko87**] claimed that he invented a similar algorithm.

This work attracted criticism from Horspool and Cormack[**Horspool87**] who claimed that they had investigated this form of algorithm and found it inferior to other, more popular techniques such as Ziv and Lempel coding. More importantly they stated that their experiments show that a climbing heuristic in which words are moved one position up the list rather than all the way to head of the list, performs as well as the LRU heuristic employed by Bentley et al. The climbing heuristic is easier to implement.

Perhaps what separates the Bentley technique from the class of Ziv and Lempel techniques is that it transmits words verbatim if they are not present in the list. Ziv and Lempel algorithms build up their dictionary implicitly and incrementally.

1.8 The Modern Paradigm of Data Compression

The algorithms that have been discussed up to this point all conform to the blocking paradigm of data compression presented in Section 1.3. Even the recent Ziv and Lempel algorithms employ blocking, though the mapping from source to channel strings changes at each step.

In the early 1980s, a new paradigm of data compression arose which we call **the modern paradigm of data compression**. The modern paradigm is provably better (in theory, with respect to compression) than blocking techniques and contains many of the previous techniques as special cases. The remainder of this chapter focuses on the modern paradigm and the compression techniques that arose from it.

1.8.1   The Information Market Place

The blocking paradigm of data compression can be likened to a market place of two traders (model/coder) who can only trade by swapping (bartering) goods (source strings/channel strings). The aim of each trader is to minimize the difference in value (information content) between the goods being traded. Unfortunately, this market of strings approach has the effect of warping the outlook of the traders to the extent that the internal organization of each becomes governed by the other. There seems little point in presenting strings of value 0.001 bits for trade if the minimum valued channel string that could be swapped for it is worth 1 bit.

In particular, the bartering mentality has impacted badly on source modelling which, up until recently, has been concerned only with choosing sets of source strings (dictionaries). An alternative to bartering is to find a currency of information that would allow the modeller to get on with modelling and the coder to get on with coding without recourse to each other.

1.8.2   Predictions: A Currency of Information

In the modern[37] paradigm, **predictions** act as a *currency* of information that allows the separation of model and coder.

In everyday life, to "predict" means to nominate an event. If the event occurs, the prediction is said to be correct. In data compression, the concept of a prediction is generalized to the specification of a finite set of exhaustive, mutually exclusive events and their probabilities. The "correctness" of such a prediction depends on the closeness with which the prediction's probabilities match the true probabilities.

The question remains of what the events of a prediction should correspond to (mean). Assumption 5 (Section 1.6) constrains the events to yield information about the remaining part of the message. The set of events could range from the set of all possible strings, to partial information about the next instance. Without loss of generality we choose the set of events to be the set of symbols. This yields the following advantages.

- Predicting a single instance is as powerful as predicting more than one instance. Once a mechanism is established which predicts

----

[37] The term "modern" has been used to describe the paradigm of data compression about to be described. For example, from [Witten87](iii): "This contrasts with the more modern model-based paradigm for coding...".

a single instance it can be invoked recursively to predict as many instances as desired.

- The instance is the smallest convenient unit of a message.

- Instance prediction data structures are relatively easy to manipulate.

We define the set of all predictions to be

$$P: \forall p \in P, \; p(A) \to \mathbf{R} \; \wedge \; \forall a \in A, \; p(a) \geq 0 \; \wedge \; \sum_{a \in A} p(a) = 1$$

Predictions are functions that map source symbols to probabilities. A prediction p is called **safe** iff $p(a) > 0$, $\forall a \in A$.

Although P is a good theoretical working definition of the set of predictions, it cannot be used in practice because a $p \in P$ is capable of containing an infinite amount of information (if, for example, $p(a)$ is transcendental for some a). In practice, each prediction must be approximated by a member of the set of all **samples**

$$X: \forall x \in X, \; x(A) \to \mathbf{Z}[0, \infty)$$

Any prediction $p \in P$ can be represented as a sample $x \in X$ with infinitesimal loss of accuracy by setting x such that $x(a) - \frac{1}{2} \leq y p(a) < x(a) + \frac{1}{2}$ where y is an abbreviation for $\sum_{a \in A} x(a)$. x is p rounded to a particular fixed-point accuracy. The representation of predictions by a group of integers not only allows a precise approximation to any p, but also allows predictions to be constructed from collections of instances sampled under particular source conditions. Each $x(a)$ stores the frequency of symbol a. Thus x is a structure that can be used in practice to turn instances into frequencies into predictions.

In summary, predictions form a currency of information. A piece of information consists of a prediction and an outcome. Under the modern paradigm, without loss of generality, the events in a prediction can be restricted to the set of symbols.

1.8.3 The Modern Paradigm

The modern paradigm uses predictions to divide compression into separate modelling and coding units. The modern paradigm is best summarized by a diagram (**Figure 9**).

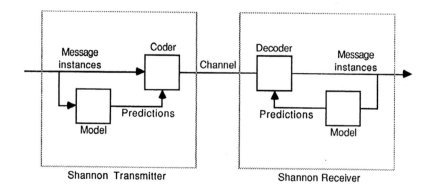

Shannon Transmitter   Shannon Receiver

The modern paradigm of data compression specifies an internal organization of Shannon's transmitter and receiver, and constrains the interface of its parts. The model units must be identical deterministic automata that receive a stream of instances and produce a stream of safe predictions, one for each instance. Each prediction must be generated before the arrival of the corresponding instance. The coder unit must accept a stream of instances and predictions of instances and use the predictions to code the instances efficiently. The decoder unit must reverse this using the identical stream of predictions generated by the receiver's model.

Figure 9: The modern paradigm of data compression.

Model: The model accepts a sequence of instances and produces a stream of predictions, one for each instance. Each prediction is produced just before the instance it predicts is read.

Coder: The coder accepts a stream of predictions and a stream of instances and produces a stream of channel instances.

Each step transmits an instance.[38] At the start of each step, the model constructs a prediction p (of the next instance) and passes it to the coder. The coder uses the prediction to transmit the next instance a using as close to $-\ln p(a)$ nats as it can. Meanwhile, the receiver's model has generated an identical prediction which the decoder uses to identify the instance that was transmitted. The transmitter and receiver both use the new instance to update their models. The cycle repeats until the entire message is transmitted.

---

[38] In fact, because of coder buffering, the channel instance(s) containing the coded instance may not be delivered to the receiver instantaneously. However, so long as the decoder can decode the instances in the order in which they were coded, the instance can be considered to be transmitted instantaneously.

Important features of the modern paradigm are listed below.

- The model (predictor) can be *any* deterministic automata so long as it produces safe predictions in finite time. However, its predictions must approximate the true probabilities if compression is to take place.

- The prediction for each message instance can be different.

The modern paradigm was put together for the first time in a landmark paper by Rissanen and Langdon[Rissanen81], who divided compression into modelling and coding. Their main theorem showed that only one instance need be predicted and is worth repeating here.

"For every adaptive or nonadaptive recursive model using alphabet extension, there exists another adaptive or nonadaptive recursive model respectively, using no alphabet extension, which has the same number of parameters (and hence requiring the same number of binary digits for its description), and which has the same ideal codelength $-\log P(s)$ for every string s. The converse is not true." [Rissanen81](p. 18(V.(ii)))

To assist in proving this theorem (for all *finite* strings), the authors defined sources as producers of finite strings. This can be contrasted to classical sources which produce infinite strings.

Cleary and Witten's theorem (Section 1.6) showing that adaptive codes are at least as powerful as static and semi-adaptive codes, in conjunction with Rissanen and Langdon's theorem showing the superiority of single-instance prediction over alphabet extension, establishes the modern paradigm as superior, in terms of compression, to other classes of algorithm.

1.8.4 Modelling

A model conforming to the modern paradigm must produce a prediction for each instance in the message. Each prediction is based upon the history. The set of all models is thus defined as

$$C: \forall c \in C, \ c(S) \to P$$

A close inspection of this definition reveals that it also serves to define the set of sources. Without loss of generality, a source can also be defined as a generator of predictions. For both models and sources, predictions are used as a method of expressing what the source is going to do next. *Thus the set C describes the set of all sources as well as the set of all models.* The incremental construction of a model can be viewed as the *reconstruction of the source* from the history.

Because real-world sources are extremely complex, it is impractical to reconstruct them exactly. Instead, it is usual to restrict consideration to a class of sources of a particular complexity. Modelling then consists of using the history to select a particular model from the class. Thus *the class of models that a compression algorithm is capable of constructing will determine the class of sources that it is capable of compressing.*

The actual class of sources selected for modelling depends on the sophistication of modelling algorithms and the amount of processing power available. Modelling is an open-ended task because its complexity is bounded only by the complexity of real-world sources; as long as the need for greater compression remains, researchers will continue to develop increasingly sophisticated models. Eventually the models will incorporate mechanisms of artificial intelligence.[39] Future data compressors might excel at compressing newspaper articles because of their knowledge of world politics.

To date, the most successful models are Markov models. Because Markov models are central to this thesis, discussion of them is deferred until Section 1.10 where they are discussed in detail.

1.8.5 Coding

The modern paradigm requires a coder with the following properties.

- Instances must be coded and decoded in the same order.

- Instances of symbols with probability $p(a)$ must be coded in as close to $-\ln p(a)$ nats of channel instances as desired.

This may seem a tall order but without such a coder, alphabet extension would prevail over single instance predictions. Luckily, the recently developed technique of arithmetic coding satisfies these conditions. Discussion of arithmetic coding is deferred until Section 1.9.

---

[39] The foundation for this development is already being layed in the field of Artificial Intelligence. For example, Dietterich and Michalski[**Dietterich85**] describe an AI technique for predicting complex sequences.

1.8.6 Not So Modern A Paradigm

The modern paradigm of data compression consists of four main concepts: the separation of modelling and coding, adaptive modelling, single-instance prediction and arithmetic coding. Each of these concepts was present in Shannon's original papers ([Shannon48][**Shannon51**]). The separation of modelling and coding is implicit in Shannon's concept of entropy. Shannon advocated single-instance prediction. The Shannon-Fano code employs the range division concept of arithmetic coding. A later paper by Shannon covers adaptive modelling and contains a diagram (**Figure 10**) that is almost identical to **Figure 9**.

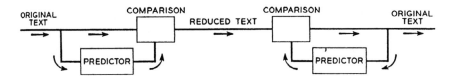

Figure 10: Shannon's compression paradigm.
(Reproduced photographically from figure 2 of [Shannon51].
Copyright © AT&T. Reprinted with special permission.)

Shannon's predictions were mere ordering of symbols. Shannon used the metaphor of identical twins to describe the concept of mutually-tracking deterministic predictors.

> "To put this another way, the reduced text can be considered to be an encoded form of the original, the result of passing the original text through a reversible transducer. In fact, a communication system could be constructed in which only the reduced text is transmitted from one point to the other. This could be set up as shown in Fig. 2, with two identical prediction devices." [Shannon51] (p. 55)

Shannon used this technique to measure the redundancy of English by asking human subjects to predict the next character of an English text basing their prediction on zero or more previous characters. The success rate of the humans gave an indication of the amount of information contained within the text. For a 100 character context, the entropy was between 0.6 bits and 1.3 bits. In 1976, Kauffman[**Kauffman76**] continued this work in the context of learning by using this method as a measure of *subjective* information in a text. The work was sponsored by the U.S. Air Force who wanted to improve their teaching programs.

Despite the fact that all these concepts were present, it took thirty years before they were combined into a coherent whole. The failure of these concepts to be integrated and adopted at the time they were discovered can be attributed to their impracticality in the 1950s (an era in which programmers spent weeks shaving off milliseconds) combined with the early adoption of Huffman coding.

1.9   Arithmetic Coding

This section presents an established coding technique called **arithmetic coding** that satisfies the coder requirements of the modern paradigm. Under the modern paradigm, the coder's task is to accept a stream of instances and a stream of corresponding predictions and code each instance in as close to the predicted entropy of the instance's symbol as possible. For example, if a prediction specifies that symbol a will occur with a probability of 0.25 then it is the coder's job to ensure that if an instance of the symbol does occur, it will use up two bits ($-\log_2 0.25$ bits) of code string. Similarly, if the prediction specified a probability of $345/999$ then it is the coder's job to code instances of that symbol in as close to $-\log_2 345/999$ bits as possible.

The technique of arithmetic coding achieves this seemingly impossible task by packing *more than one source instance* into the same channel instance. The following description, which differs from the usual expositions of arithmetic coding, follows the reasoning that led the author of this thesis to belatedly rediscover the technique in late 1986; it is closest to Guazzo's approach[Guazzo80].

1.9.1   A Description of Arithmetic Coding

In this approach, coding is viewed as the *filling of buckets* (Section 1.3.1). The coder receives packets of information which are placed into buckets. When a bucket is full, it is shipped to the channel.

In the blocking paradigm, recognition of a source string implies the emission of the corresponding channel string. Coding and decoding is simple because the events are constrained to correspond to codes that are multiples of the channel bucket capacity. In the modern paradigm, the coder must code packets containing an *arbitrary* amount of information. If coding is to be efficient, a technique must be found for *partially filling* a bucket (**Figure 11**).

In fact partial bucket filling is commonplace. Consider a coder about to fill a bucket of size 256 with the outcome of an event that succeeds or

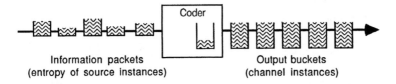

Information packets   Output buckets
(entropy of source instances)   (channel instances)

In the past, coders could only efficiently code source events whose information content was a multiple of the bucket capacity. In contrast, arithmetic coders can accept a stream of events of arbitrary information content (including events containing far less than a bucketful of information) and code them as efficiently as required. Logically, the arithmetic coder treats the information as a liquid (instead of a solid), using the arriving information to fill each bucket exactly before shipping it.

Figure 11: The coder's job is to fill output buckets.

fails with equal probability. If the bucket is viewed as an eight bit byte, the coder can optimally code the outcome by setting the "top bit" of the byte to 0 upon failure and 1 upon success. This can be done by adding either 0 or 128 to the byte. The fact that this information can be placed in the bucket *arithmetically* conflicts with the notion of the bucket being indivisible. What is happening here?

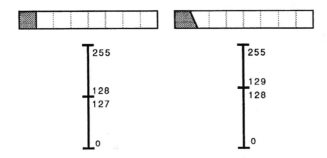

Traditionally, a byte is filled with information by setting its bits in turn (left top). The setting of the top bit of a byte can be viewed as the halving of the range of the byte (left bottom). By dividing the range unevenly, arithmetic coders can efficiently code events of arbitrary probability. The range division shown on the right could be used to accurately code two events of probability 127/256 and 129/256.

Figure 12: Restricting ranges is the same as filling bits.

If the top bit is 0, the value of the byte is in the range $\mathbf{Z}[0, 127]$; if 1, the value of the byte is in the range $\mathbf{Z}[128, 255]$. Each outcome is allocated half the *range* of the byte. This division serves to represent exactly one bit because 128/256 of the byte's possible values are allocated to each event of probability 1/2. Now consider the case in which the probabilities are not 1/2 each but rather 129/256 and 127/256. This event can be optimally coded by the same splitting process, the only difference being that we can no longer point to a "bit" (**Figure 12**). Constraining a number from a range of width a down to a range of width b corresponds to the transmission of $\ln a - \ln b$ nats. For large a and b with $a \approx b$, this information quantity can be very small. Arbitrary range divisions can be decoded by the decoder in much the same way that bit aligned divisions can be decoded.

Returning to the example, so far a single event has been coded into a bucket by splitting the bucket's range into two parts and using the outcome of the event to select one of the ranges. If the bucket were shipped at this stage, it would contain roughly one bit of information and most of its capacity would be wasted. The remainder of the bucket's capacity can be used by treating the remaining range as another bucket. If the event with probability 129/256 occurred, there will remain a "bucket" of size 129. If the event with probability 127/256 occurred, there will remain a "bucket" of size 127. This range can be divided again and the process repeated.

Eventually, the range decreases to a single value (the bucket fills up). When this happens the bucket must be shipped and a new (empty) bucket used.

The description above covers the basic mechanism of arithmetic coding. However, some tricky implementation problems must be resolved before the technique becomes practical.

Problem: Range Resolution Coding the events of probability 127/256 and 129/256 onto the range $\mathbf{Z}[0, 255]$ was efficient because the range had exactly enough resolution to represent the division. As a bucket fills and its remaining capacity decreases, so does the accuracy with which the remaining range can accommodate a range division. In the extreme, the coder might be expected to code a prediction of $(1/1000, 999/1000)$ into a range $\mathbf{Z}[0, 1]$. In such a case, the coder would have to allocate the range $\mathbf{Z}[0, 0]$ to one symbol and the range $\mathbf{Z}[1, 1]$ to the other resulting in a great inefficiency (because the 999/1000 event contains far less than one bit of information). If there were three symbols to code, the coder would be blocked.

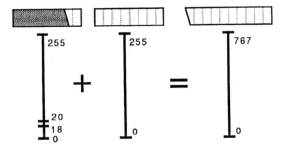

Two buckets of different sizes (e.g. a partially full and an empty
bucket) can be treated as one large bucket whose range is the product
of the ranges of the two smaller buckets.

Figure 13: Addition of a new bucket extends the range.

Whenever the range of the bucket being filled becomes too narrow
to support the level of accuracy required to efficiently code the next
prediction, a new empty bucket can be used to widen the range out
again. The new bucket is merged logically to the old bucket and the
two buckets treated as a *single* super bucket. The new super bucket has
a size that is the *product* of the remaining sizes of the old, nearly-full
bucket and the new empty bucket. For example, if the old bucket had
a remaining range of $Z[18, 20]$ (a size of 3) and the new bucket had a
range of $Z[0, 255]$, the new super bucket would be of size 3×256 (size
768) (**Figure 13**). A good way of viewing the super bucket is as a two
digit number with the first digit expressed in base 3 and the second in
base 256. The two buckets together form a new range that can be divided
in the original manner. Eventually, the divisions result in the range of
the first bucket diminishing to a single value. When this happens, that
bucket can be shipped.

The technique of bucket merging can be performed with many
buckets. By adding new buckets whenever the size of the super
bucket falls below a threshold, any desired accuracy can be maintained
indefinitely. By shipping full buckets, only finite memory is required to
maintain the accuracy. If the bucket size is small (e.g. 2), many buckets
can be joined together to form a target range of the required accuracy.

Problem: Carry-Over Maintenance of a finite, bounded number of
buckets in a super bucket "filling station" is possible provided that full
buckets can be shipped when new buckets are created. Unfortunately,
such a property is not guaranteed, as it is possible for the division of
the super bucket's range to occur in such a way that the top digit of the

super bucket (the value of the oldest bucket) is never resolved. In such circumstances, the oldest bucket cannot be shipped. This results in a build up of nearly-full buckets.

The author of this thesis couldn't solve this problem. Nor, it seems could Guazzo.

> "One disadvantage of the algorithm is that we have to keep a buffer of secondary symbols that are still liable to be overwritten. This should be no surprise; if we are unwilling to send to the output symbols that are not "full" of information, we should be prepared to wait until we have some "piece of information" that fits into them." [Guazzo80](p. 19)

In a strict sense, Guazzo is correct. However, in practice, the buffer itself can be compressed to $O(\log n)$ of its previous size, using negligible space. A close examination shows that deferment of bucket shipping can only occur when all the deferred buckets (except for the working buckets and one other bucket) have the maximum value. This means that buffered buckets can be stored by a counter. It is poetic justice that arithmetic coding algorithms use run-length coding internally. Other solutions also exist (Section 1.9.2).

The "derivation" of arithmetic coding given above was motivated by practical considerations. In contrast, other presentations of the technique suggest that every other discovery of the technique was motivated by theoretical considerations. In *theory*, arithmetic coding is even simpler than might be supposed from the discussion above. If the multi-bucket register is considered to be of infinite length, the technique becomes entirely arithmetic with no bucket manipulations involved. The principle behind the theoretical technique is simply:

Principle of Arithmetic Coding: A set of messages is optimally coded by exhaustively dividing the range $[0, 1)$ among the messages in proportion to their probability. A message is coded by transmitting any number in its subrange using channel digits (The narrower the range, the more digits being required).

Thus, in theory, an arithmetic code would represent a ten megabyte message as a single, very accurate number in the range $[0, 1)$. As numbers of such accuracy are intractable to manipulate, practical arithmetic coding relies on using fixed precision arithmetic, which boils down to bucket shuffling. In practice, buckets are not manipulated explicitly. Instead, an arithmetic code manipulates two fixed width registers called the A register and the C register (from [**Langdon84**]). The A register stores the width of the current range. The C register stores the position of the lower bound of the current range. The two registers can be

imagined to be positioned at the end of the code string being generated
(**Figure 14**). Whenever the *A* register (which represents the available
precision) gets too small, both registers are shifted to the right (their
contents are shifted left). It's all done with shifting, adding and mirrors.
The result is a highly efficient, practical technique.

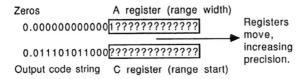

Arithmetic coding can be implemented using two registers that operate on the end of the code string. Here, coding is viewed in the more
theoretical sense as the production of a single number in the range
$R[0, 1)$. The A register holds the width of the remaining range. The
C register holds the lowest digits of the start of the range. As the
range narrows, the registers conceptually shift to the right emitting
the code string to the left.

Figure 14: Arithmetic coding using fixed precision registers.
(Based on the diagram on p. 802 of [Langdon82])

In summary, arithmetic codes operate by subdividing ranges in
proportion to the specified probabilities. Ingenious mechanisms are used
to overcome the problems of resolution maintenance and bucket buffering.
Arithmetic codes satisfy the requirements of the modern paradigm of
data compression by coding a sequence of events of continually changing
probabilities in as close to their information theoretic space as desired.

1.9.2 The Development of Arithmetic Coding

Although the idea behind arithmetic coding is simple, the field of
arithmetic codes is messy and complex. There is no single code called
"the arithmetic code". Instead there is a class of codes. The development
of arithmetic codes has been messy too; no brilliant flash ushering in the
new age; rather, three decades of sparks which finally ignited in a few
key papers. This section does not aim to unravel these developments,
only outline them. For a more detailed description see Appendix 1 of
[Langdon84].

There are a variety of arithmetic codes and many different ways of
implementing them. The technique of practical arithmetic coding can be
decomposed into three sufficient concepts:

- the representation of a message by a fraction in the range $[0, 1)$.

- the use of fixed-precision registers.

- the avoidance of indefinite carry over.

The idea of a message being represented by a fraction first appeared in Shannon's 1948 paper[Shannon48](p. 402) as part of the proof of his "Fundamental Theorem for a Noiseless Channel". Shannon proposed that messages be arranged in decreasing order of probability and that the chosen message be transmitted as a binary fraction (being the sum of the probabilities of messages earlier in the list) long enough to identify the message's probability range. This idea is used in the Shannon-Fano code described in Section 1.4.1.

It took thirty years before Shannon's idea was developed into a practical coding scheme. During this time, the area received attention by Abramson, Elias, Jelinek, Schalkwijk, Cover (see [Langdon84] for references) but was blocked by the twin problems of precision and the requirement of having to decode the instances in the same order in which they were encoded (FIFO (First In, First Out) operation as opposed to LIFO (Last In, First Out) operation).

In 1976, Pasco[**Pasco76**] and Rissanen[**Rissanen76**] independently[40] discovered the use of fixed-length registers. However, as presented, both techniques were impractical. Rissanen's technique was LIFO and Pasco's, while FIFO, did not solve the indefinite carry over problem.

Another burst of publications followed in 1979, detailing practical techniques. Guazzo[Guazzo80] described a practical arithmetic coding scheme but couldn't solve the carry over problem (Section 1.9.1). Guazzo showed that arithmetic coding allowed a source with one set of characteristics to be coded into a channel with a different set of characteristics (a **constrained channel**) including the case of a channel whose symbols occur with equal probability. Rubin[**Rubin79**] refined Pasco's FIFO algorithm, by describing a technique for preventing carry over. Rissanen and Langdon[Rissanen79] not only described a practical technique but generalized and characterized the family of arithmetic codes.

Three solutions to the carry over problem have been proposed. The first, proposed by Guazzo is simply to terminate the code (as if the end of the message has been reached) every so often and continue afresh.

---

[40] Actually, whether the discoveries were independent is unclear. Appendix 1 of [Langdon84] implies that they were independent whereas the second paragraph of [Rissanen79] states that they were not.

The second solution, proposed by [**Jones81**](IV.iii(22–24)), is to use a counter to store the buckets that threaten to overflow (i.e. internal run-length coding). The third solution, called **bit stuffing**, by Langdon and Rissanen[Langdon81], is to insert a 0 bit into the output stream after a fixed limit of 1s have been shifted out of the coder register; the artificially placed 0 bit catches any carry that might occur and allows the run of 1 bits to be written immediately. For a limit of k bits, an extra bit will be inserted on average once every 2^k output bits (because the output approximates a memoryless binary source with equiprobable symbols).

Of these solutions, bit stuffing is probably the best in practice. Regular termination of the message is inelegant and wastes space. The use of a counter is extremely simple and elegant but does not guarantee an upper bound on the effective buffering (a disadvantage for real time systems); if the code string is 1111111... (for a binary channel), the first instance of the code string will appear only after the entire message has been processed. Bit stuffing does not require regular termination and provides an upper bound on the effective buffering.

By 1980, arithmetic coding had been well fleshed out. It remained only to popularize it and to refine its implementation.

1.9.3 The Popularization of Arithmetic Coding

Although arithmetic coding was well-developed in 1980, there was no clear public image of it as there was (say) of Huffman coding. The situation is much the same today despite the publication of "accessible" descriptions.

In 1981, Langdon wrote a comprehensive "Introduction to Arithmetic Codes" in an IBM Research Report. This was revised and republished in 1984[Langdon84]. Jones[Jones81], who independently discovered arithmetic coding, gave a detailed description of an algorithm down to the program code level. Witten, Neal and Cleary[Witten87] probably drew the most attention to arithmetic coding through their publication in the popular *Communications of the ACM* in which they not only described the technique but also presented a full implementation in one of the most prevalent programming languages — C [**Kernighan88**]. Despite all these efforts, arithmetic coding remains relatively unknown.

1.9.4   Refinements of Arithmetic Coding

By 1980 there were arithmetic codes that could code predictions to as close to their entropy as required. Further work focussed exclusively on making the codes more efficient.

The most important refinement work on arithmetic codes was Langdon and Rissanen's work on **binary arithmetic codes**. Binary arithmetic codes operate upon binary source and channel symbol sets. Just as binary run-length codes are special, so are binary arithmetic codes. In particular, binary arithmetic codes can be used to code a source symbol set of arbitrary size by arranging the symbols in a binary tree and coding each traversal decision.

In 1981, Langdon and Rissanen[Langdon81] published the remarkable fact that for a prediction consisting of two events, the probability of the least likely event could be approximated by a power of two (i.e. $p \approx \frac{1}{2}, \frac{1}{4}, \frac{1}{8} \ldots$) with a relative loss of compression of at most 4%. This fact was used to replace the multiplication operation of previous arithmetic codes with a more efficient shift operation (at the $< 4\%$ cost) — a great improvement over previous codes. The simplified code is described by four simple equations involving only shift and addition operations. A and C denote the entire *numbers* of **Figure 14**. $Q(s)$ is the **skew number**, being minus the power of two approximating the probability of the least probable symbol. Here, a prediction consists of a skew number [**Langdon79**] and a bit identifying the least probable symbol, which for the purposes of exposition, we assume to be 1.

$$A(s0) = A(s) - A(s1)$$
$$A(s1) = A(s) \times 2^{-Q(s)}$$
$$C(s0) = C(s)$$
$$C(s1) = C(s) + A(s0)$$

In a later paper[Langdon82], an additional approximation was used to make the technique even more efficient. A minor problem with the technique above is that the subtraction cannot take place concurrently with the shift. These operations were made concurrent by replacing $A(s)$ in the second equation by the greatest power of two less than or equal to $A(s)$. This resulted in an even simpler algorithm suitable for hardware implementation.

In a joint effort between different IBM research centers, the code in [Langdon82] was developed into a general purpose "turnkey" binary arithmetic code called the **Adaptive Q-Coder (or Q-Coder for** short) which is suitable for hardware and software implementation[**Langdon88**][**Mitchell87**]. In particular, algorithm "[Langdon82]" was modified to improve compression performance by replacing the powers of two ($2^{-Q(s)}$ values) in the equation by precomputed values slightly larger than the greatest power of two less than or equal to $A(s)$. These values were precomputed and stored in a lookup table. This modification partly compensated for the approximations made in [Langdon82].

The Q-coder also keeps track of the frequencies of the binary symbols and uses them to make the prediction (taking the form of most probable symbol and a power of two) for each message instance. The user merely feeds the Q-coder a sequence of binary events and it does the rest. Further discussion of the Q-coder is deferred until Section 1.11.7.

Martin, Langdon and Todd[**Martin83**] adapted arithmetic codes for use on constrained channels such as magnetic recording devices in which state transitions cannot occur too often or too rarely.

1.10 Markov Models

The modern paradigm permits the use of any model that can produce a stream of safe predictions, one for each instance. Currently, the most successful class of model is the class of Markov models, particularly the subclass of variable-order, finite-context Markov models. Later chapters of this thesis focus on Markov models and use them as a basis for exploring adaptive techniques. This section discusses Markov models in detail.

1.10.1 Markov Chains, Markov Sources and Markov Models

In this section we give a brief description of Markov chains[**Feller57**] [**Karlin69**][**Bhat72**] and their relationship to Markov sources and Markov models.

A **Markov chain** consists of a set of **states**. The set can be infinite but we will consider only finite Markov Chains containing q states numbered 1 to q with $Q = \{\mathbf{Z}[1, q]\}$. Time is quantized. At any point in time, a chain is in a particular state. At each tick, the chain changes state (possibly to the same state). If a chain is in state i, it will change to state j with probability $P_{i,j}$. The conditions

$$\forall i, \sum_{j \in Q} P_{i,j} = 1 \quad \text{and} \quad \forall j, \sum_{i \in Q} P_{i,j} = 1$$

hold; each row forms a prediction of the next state. A Markov chain is a directed graph, each node of whose outward arcs are labelled with probabilities summing to one.

A **Markov source** is the same as a Markov chain except that each node has exactly n outward arcs, each labelled with a different symbol. At each tick, the source moves to the next state in accordance with the probabilities on the outward arcs of its current state, and emits an instance of the symbol associated with the arc taken.

Markov Models are identical to Markov Sources (Section 1.8.4). Markov sources and models are simply stochastic deterministic[41] finite state machines.

As Markov models tend to be unmanageable unless they are finite and ergodic, we add a new assumption to the list of assumptions already accumulated (Section 1.1.3 and Section 1.6).

Assumption 6: Sources are finite and ergodic.

A finite ergodic source is a source that has a finite number of states and for which the statistical properties of its output converge as the length of the output approaches infinity. The precise mathematical definition of a finite ergodic source varies from one probability text to another. Here the term is used to indicate that the source in question is a Markov chain, that it has a finite number of states, that it is possible through zero or more steps to move from any state to any other, and that the greatest common divisor of the lengths of all its cycles is one. The last restriction is necessary to ensure that the chain does not retain phase information at time infinity.

1.10.2 Constructing Models from Histories

A difficulty in constructing arbitrary Markov models is the lack of correspondence between states and symbols. Even if the model probabilities are known, the problem of determining the state of a Markov chain from a long history string (without ever being given the state of the source) is non-trivial and non-deterministic. At best, we can only obtain a probability distribution of the model's current state.

---

[41] Deterministic in the sense that no two arcs leading from any node are labelled by the same symbol.

To date, no one has found a feasible algorithm for constructing arbitrary Markov models from history strings.[42] A particular string can correspond to many models and it is difficult to arrive at a good model.

These problems can be transcended by considering only models whose state can easily be determined from the history string. In fact it is desirable to consider only those models whose state can be determined from a *constantly bounded* number of the most recent history instances. We add this assumption to the others.

Assumption 7: The state of a source can be determined by inspecting the previous m instances that it has emitted.

This assumption forces a *direct* correspondence between the internal state of a model and the string that it has produced. It also means that the entire history need not be stored in order to determine the current state.

Bell and Moffat, two researchers currently active in this area, have associated classes of finite state automata with the source models obtained with and without the above assumption[**Bell88**]. Without the assumption, the model is a **Finite State Automaton**. With the assumption, it is a **Finite Context Automaton**. The distinction is important because it allows us to consider the set of all sources as the set of mappings from strings of length m to predictions:

$$D: \forall d \in D, \; d(S_m) \rightarrow P$$

The domain of any $d \in D$ can be extended to shorter strings by defining $d(S_l)$, $\forall l \in \mathbf{Z}[0, m-1]$ in terms of d's implicit steady state probabilities. The set of sources D is called the set of **order-m finite-context Markov sources**.

In practice, each prediction $d(s)$ ($s \in S_m$) in each source d in the set D can be approximated by a sample x_s (Section 1.8.2). The dual nature of samples suggests a method for constructing a sample/prediction x from a history string h of length l. $\boxplus$ returns the number of times that its argument is *true* over the specified domain.

$$\forall t \in S_m, \; x_t(a) = \biguplus_{k=1}^{l-m} (h_{k...k+m-1} = t) \wedge (h_{k+m} = a)$$

---

[42] The Forwards Backwards algorithm comes close but suffers from drawbacks discussed in Section 1.10.6.3.

This method treats the message as a statistical sample. Better still, it allows the source model to be constructed incrementally from the message string. If the source string is long enough, the probability of an instance of the symbol a following the string $s \in S_m$ in a source stream can be estimated as $x_s(a)/y_s$. This sampling technique has good asymptotic properties ($\lim_{l \to \infty} x_s(a)/y_s = d(s)(a)$) but does not produce safe predictions. To produce safe predictions, a non-obvious estimation technique must be used. The problem of finding such a technique is central to practical data compression and justifies a detailed discussion.

1.10.3   Estimation Techniques

We define the set of all estimation techniques Ξ to be the set of all mappings from samples to predictions.

$$\forall \xi \in \Xi, \ \xi(X) \to P$$

Of these estimation techniques, only those that satisfy the following two requirements are of any practical interest.

The first requirement is the estimation technique generate only safe predictions. An estimation technique ξ is of interest iff

$$\forall x \in X, \ \forall a \in A, \ \xi(x)(a) > 0$$

This problem of producing safe predictions in the face of zero frequencies is called the **zero frequency problem**[Roberts82].

The second requirement is that the estimation technique's predictions converge on the samples as the sample size increases. An estimation technique ξ is of interest iff

$$\forall x \in X \colon y > 0, \ \forall a \in A, \ \lim_{k \to \infty} \xi(kx)(a) = x(a)/y$$

where kx denotes the sample x with each frequency multiplied by k.

The assignment of positive probabilities to symbols that have a zero frequency may appear to be an action motivated entirely by the practical need for safe predictions. In fact, this assignment is supported by statistical theory as well. When a sample x of y instances is taken from an infinite population described by the prediction p, each $p(a)$ has to manifest itself in an $x(a)$ as a member of the set $\{\frac{0}{y}, \frac{1}{y}, \frac{2}{y}, \dots, \frac{y}{y}\}$. This leads to problems if a $p(a)$ is less than $1/y$. For large y, the law of large numbers states that on average, each $p(a)$ will manifest itself in the $x(a)$ value that minimizes $|x(a)/y - p(a)|$. By inverting this mapping we can, for a given y, map sample frequencies to probability ranges to the means of the ranges.

$$x(a) = 0 \quad\Longrightarrow\quad p(a) \in [0, \tfrac{1}{2y}) \quad\Longrightarrow\quad \overline{p}(a) = \tfrac{1}{4y}$$

$$0 < x(a) < y \quad\Longrightarrow\quad p(a) \in [\tfrac{2x(a)-1}{2y}, \tfrac{2x(a)+1}{2y}) \quad\Longrightarrow\quad \overline{p}(a) = \tfrac{x(a)}{y}$$

$$x(a) = y \quad\Longrightarrow\quad p(a) \in [\tfrac{2x(a)-1}{2y}, 1] \quad\Longrightarrow\quad \overline{p}(a) = \tfrac{4y-1}{4y}$$

The fact that each mean is positive indicates that there is a theoretical basis for assigning small, positive probabilities to symbols of zero frequency. The question of what the values should be is one that has baffled philosophers and statisticians for hundreds of years. It is generally considered that no objective solution exists[**Witten86**]. In practice, any "small" positive value will suffice. Most practitioners seem to find making this arbitrary decision distasteful.

Common estimation techniques form two groups: linear and non-linear.

1.10.3.1 Linear Estimation

Linear estimation techniques allocate a small piece of probability to all n symbols and then divide the remaining probability between the symbols of positive frequency. The following generalized linear formula does this for symbol $a \in A$ and coefficients $b, c, d, e \in \mathbf{R}(0, \infty)$.

$$\xi(x)(a) = \frac{bx(a) + c}{dy + e}$$

In fact, the four coefficients can be reduced to one by noting that the probabilities of a prediction must sum to one.

$$\sum_{a \in A} \xi(x)(a) = 1 \qquad \text{so} \qquad \sum_{a \in A} \frac{bx(a) + c}{dy + e} = 1$$

$$dy + e = \sum_{a \in A} bx(a) + c$$

$$\frac{dy + e - nc}{b} = y$$

$$dy + e - nc = yb$$

Because y is a free variable, we can separate into two constraint equations:

$$b = d \qquad \text{and} \qquad e = nc$$

Thus the only form of the first equation that will produce sensible probabilities is

$$\xi(x)(a) = \frac{bx(a) + c}{by + nc}$$

which can be normalized by dividing by b

$$\xi(x)(a) = \frac{x(a) + c/b}{y + nc/b}$$

b and c can be collapsed into a single parameter $\lambda = nc/b$. This yields a normalized linear form

$$\xi(x)(a) = \frac{x(a) + \lambda/n}{y + \lambda}$$

The $\lambda \in \mathbf{R}(0, \infty)$ parameter allows the specification of a greater or lesser period of transition from uniform probabilities ($\xi(x)(a) = 1/n$) to those of the sample ($\xi(x)(a) = x(a)/y$). The smaller the value of λ, the greater the trust in the sample.

It is possible to show that if all possible probability distributions are equally likely, the optimal estimation formula is linear estimation with $\lambda = n$. Appendix A contains a derivation by Jones showing this. In practice the optimal λ seems to be about 1 (Experiment 2, Section 4.17.5).

1.10.3.2 Non-Linear Estimation

In non-linear estimation, symbols of zero frequency are treated as a special case. The available probability is divided into two parts, one part of which is divided evenly among the symbols of zero frequency and the other part of which is divided among the remaining symbols in proportion to their frequency. Here, dividing the probability within each group is easy because there is a natural solution in each case. The difficulty is splitting the probability between the two groups. As before, this question has no objective solution and we parameterize in λ using the same argument developed for linear estimation.

$$\xi(x)(a) = \begin{cases} x(a) = 0 & \to \frac{1}{z}\frac{\lambda}{y+\lambda} \\ x(a) > 0 & \to \frac{x(a)}{y}\frac{y}{y+\lambda} \end{cases}$$

$z \in \mathbf{Z}[0, n]$ is the number of symbols of zero frequency in x ($z = \biguplus_{a \in A} x(a) = 0$).

1.10.3.3 Linear and Non-Linear Moffat Estimation

In the linear and non-linear estimation formulae, the λ parameter acts as an estimate of the likelihood of the arrival of an instance of a symbol of zero frequency. The higher the value of λ, the more the formula is betting on the occurrence of a new symbol.

The difficulty with choosing a fixed value for λ is that different values are appropriate for distributions of different entropy. For flat distributions high values of λ are appropriate; for spiky distributions low values work best.

In 1988, Moffat[**Moffat88**] described a non-linear technique that sets λ to $n-z$ dynamically.[43] By setting $\lambda \leftarrow n-z$, the technique effectively counts the occurrence of **novel**[44] symbols. This means that λ is regulated in accordance with the spikiness of the distribution.

Moffat calls this estimation technique "Method C" (after methods A and B in [**Cleary84**]). We will refer to it as **nonlinear Moffat estimation** and generalize it so that it contains a λ value as the other estimation techniques do. Moffat's technique corresponds to the special case $\lambda = 1$.

$$\xi(x)(a) = \begin{cases} x(a) = 0 & \rightarrow \dfrac{1}{z}\dfrac{\lambda(n-z)}{y+\lambda(n-z)} \\ x(a) > 0 & \rightarrow \dfrac{x(a)}{y}\dfrac{y}{y+\lambda(n-z)} \end{cases}$$

A similar modification can be made to the linear formula, yielding **linear Moffat estimation**.

$$\xi(x)(a) = \frac{x(a) + (n-z)\lambda/n}{y + (n-z)\lambda}$$

1.10.4 Summary of Fixed-Order Markov Methods

Before moving onto variable-order methods, it is worth summarizing the fixed-order Markov methods outlined in the previous few sections.

The designer chooses an **order** $m \in \mathbf{Z}[0,\infty)$ (usually $m \leq 4$). The order determines the number of instances that are used to predict the next instance. The order m defines n^m contexts (called **conditioning classes** in Rissanen/Langdon[**Rissanen81**] terminology) which are uniquely identified by members of the set S_m. The model consists of a group of samples $x_s \in X$, one for each context $s \in S_m$. At the start

---

[43] If $n = z$ the value $\lambda = 1$ is used.
[44] The term "novel" is often used to describe symbols of zero frequency.

of transmission, all samples are set to empty ($x_s(a) \leftarrow 0$ for all $s \in S_m$ and $a \in A$). Before each instance is read, the context x_s is used to make a safe prediction $\xi(x_s)$ where $s = h_{|h|-m+1...|h|}$ (the most recent m instances in the history) (ξ is chosen by the algorithm designer from the estimation techniques discussed in Section 1.10.3). After the new instance a is transmitted, $x_s(a)$ is incremented. The process repeats for as long as required.

This technique is called a **fixed-order technique** because it bases each prediction on the same number (m) of history instances. Each instance is predicted based on the instances that occurred in similar contexts of length m in the history.

Example: A zero-order ($m = 0$) technique uses a single (n^0) context that corresponds to the empty string ϵ. The context records every instance in the history string ($x_\epsilon(a) = \biguplus_{i=1}^{l} h_i = a$).

Example: A first-order ($m = 1$) technique uses n contexts (n^1) that correspond to the n symbols. Each context stores the frequency of instances of different symbols following instances of the context's symbol in the history ($x_b(a) = \biguplus_{i=1}^{l-1} h_{i...i+1} = ba$ ($b \in S_1, a \in A$)).

Example: A second-order ($m = 2$) technique uses n^2 contexts corresponding to the n^2 symbol pairs. Each context stores the frequency of instances of different symbols occurring after instances of the context's symbol pair in the history ($x_b(a) = \biguplus_{i=1}^{l-2} h_{i...i+2} = ba$ ($b \in S_2, a \in A$)).

1.10.5 Variable-Order Techniques

Fixed-order models are simple but suffer from two problems that make them impractical except for very low m.

Memory Consumption: An order m model requires memory for n^m samples, each of which contains n counters. Using today's computer storage technology with $n \simeq 256$, fixed-order methods become impractical at $m \simeq 2$. One optimization is to use a sparse representation, storing only contexts that contain one or more instances. This reduces the memory requirement to the smaller of n^m and l contexts, where l is the length of the message.

Sample Significance: The second difficulty is the trade-off in speed and accuracy between higher and lower order models. Consider an order 0 model containing one context and an order 100 model containing n^{100} contexts. For a history of length l, the order zero model will have placed the l instances into its only sample and will make a prediction based on a sample of l instances. On the other hand, the order 100 model will have distributed the l instances amongst its n^{100} samples and will make a prediction based on an average sample of l/n^{100} instances which for reasonable length messages ($l \ll n^{100}$), will be of zero size. Thus for $l \ll n^{100}$, the order 0 model will perform better than the order 100 one. It is guaranteed, however, that as $l \rightarrow \infty$, the order 100 model will perform at least as well as the order zero model. Thus it is better to use lower order models at the start of a message and switch to higher order models as samples fill.

Variable order techniques avoid the learning problem by maintaining models of many orders at the same time. Predictions are based on a combination of the samples of the different ordered models. Variable order techniques avoid the memory problem by organizing the contexts of the different orders of model into a single tree structure that can be pruned to suit the memory requirement.

1.10.6 An Overview of Markov Algorithms

This section contains a survey of the variable-order, finite-context Markov algorithms that have appeared to date. The term "algorithm" is used here to describe researchers' efforts where the word "model" is probably more appropriate. However, because researchers tend to describe their compressors as algorithms rather than models, the word "algorithm" is preferred.

Before launching into a discussion of Markov algorithms, a historical note on nomenclature is in order. As far as the author can tell, each of the algorithms to be presented was invented independently of the others. This has meant that each algorithm has been named after the general technique that it uses (e.g. Variable Order Markov Modelling) rather than the characteristics that distinguish it from similar techniques. Because any sensible description of Markov modelling requires a few long words, the result is names such as "Local Order Estimating Markovian Analysis", "Prediction by Partial Matching", "Dynamic History Prediction Compression" and "Dynamic Markov Compression" all of which basically describe the same thing.[45] The way that the field has recovered from this historical mess has been to acronymize all the names and pretend that the acronyms really stand for names

---

[45] By this is meant not that the objects that they denote are the same, but that the names themselves, when viewed in isolation, all describe the characteristics of the class of algorithm to which the objects all belong.

that distinguish one algorithm from another. This fragile tradition is perpetuated in this thesis.

1.10.6.1 Markov Sources by Shannon

Shannon spotted the relationship between Markov chains and Markov sources in his founding paper[Shannon48]. Shannon modelled sources as stochastic processes that generate instances one at a time, and described a series of finite-context, fixed-order Markov sources of increasing order as an approximation to English. He did not tackle arbitrary Markov sources.

In a later paper, Shannon[Shannon51] proposed a compression system (involving prediction based on n-gram frequencies) that embodied many of the concepts of the modern paradigm (Section 1.8.6). Shannon's predictions consisted not of probabilities but of orderings on the set of symbols.

Shannon's technique was not adopted generally, presumably because at the time it would have been expensive in memory and processor time.

1.10.6.2 DAFC

The simplest variable-order algorithms maintain a zero order model and a first order model. These algorithms are best described as **fractional-order algorithms**. Whereas zero order models use one context (which collects instances of all symbols), and first order models use n contexts (each of which collects instances of a particular symbol), a fractional order model uses $k \in [2, n-1]$ contexts (each of which collects instances of one or more symbols).

The problem of forming a many-to-one relation between n symbols and k contexts has n^k solutions — less if each context has to correspond to at least one symbol. The most skewed allocation allocates $k - 1$ symbols to separate contexts and the remaining $n - k + 1$ symbols to the remaining context. The most uniform allocation allocates approximately n/k symbols to each context. Both extremes have been investigated.

The technique of allocating a small number of first-order contexts among symbols is not a new one (Section 1.4.3). However, earlier techniques were static (e.g. [Cormack85]), requiring that the designer perform the allocation.

In 1983, Langdon and Rissanen[**Langdon83**] described a fractional-order algorithm called DAFC (Double-Adaptive File Compression) that *dynamically* allocates symbols to 32 contexts. Initially all symbols are lumped into a single context (numbered 0). As instances of the symbols arrive, the frequency count of each symbol increases. The first 31 symbols whose frequency exceeds a constant threshold (the paper recommends a threshold of 50) are allocated a context (numbered $Z[1, 31]$). Once a context is allocated, it remains allocated forever. The motivation for allocating contexts to the most frequent symbols is to increase the proportion of time in which the model is generating first-order (as opposed to zero-order) predictions. The technique was augmented with run-length coding. Coding is performed by decomposition using the simple binary arithmetic code ([Langdon82], Section 1.9.4). As might be expected, the algorithm yields compression between that of a zero and first order model and consumes memory between that of a zero and first order model.

The other extreme has been investigated by Jones[**Jones88**] who simply mapped n/k arbitrary (a modulo method was used) symbols to each context. He then plotted a graph of compression vs k and obtained a fairly smooth increase of compression performance with increasing k.

Without exception, the variable-order Markov models to be described allocate one context to each member of a group of privileged symbols and lump the remaining symbols into a single context.

1.10.6.3 FBA

The Forward-Backward algorithm (FBA), described in Roberts's Ph.D. thesis[Roberts82], is the only algorithm described here that does not make Assumption 7 (Section 1.10.2). The FBA algorithm accepts a Markov model with a priori transition probabilities. It then scans the message repeatedly, each time modifying the transition probabilities so as to increase the probability of the model generating the string if left to run generatively in equilibrium.

Roberts reports that a data compression algorithm based on the FBA yielded 4.8 bits/word for "the laser patent text" which is a concatenation of patents relating to lasers. Although this algorithm is quite powerful, it has some severe disadvantages.

- It needs to scan the message repeatedly. In order to use it under the modern paradigm, it would have to be operated in a block-wise fashion.

- It is expensive in memory and processing time.

- It does not guarantee to find the optimum probabilities; it uses what is effectively a hill climbing technique.

- It is prone to "overparameterization", a condition that occurs when a model adapts itself more to the message than to the source.

These disadvantages make the FBA unsuitable for data compression.

1.10.6.4 LOEMA

The LOEMA (Local Order Estimating Markov Analysis) algorithm by Roberts[Roberts82] was the earliest use of variable-order, finite-context Markov models for data compression. The aim of the work was to overcome the two problems with fixed-order models that were described in Section 1.10.5.

Although it is not explicitly specified, LOEMA appears to organize its contexts in a backwards tree which is stored in a hash table. Predictions are made by blending the predictions of matching contexts. The blending weightings are calculated from the confidence value $C_i \in \mathbf{R}[0,1]$ of each prediction (of order i). The final prediction p predicts symbol a with probability

$$p(a) = C_m p_m(a) + (1 - C_{m-1})(C_{m-2}p_{m-2}(a) + (1 - C_{m-2})$$
$$(Cm - 3p_{m-3}(a) + (1 - C_{m-3})(\ldots(p_0(a))))$$

where p_i is the prediction of the model of order i.

One of Roberts's aims was to conserve memory. Roberts discusses two ways of restricting the growth of the Markov tree. The first is to grow leaves and then prune those that do not perform well. The second is to use a metric of leaf worthiness to determine where to grow leaves. Roberts used the product of the entropy advantage of the presence of the parent node in the tree and the entropy of the parent node's prediction. The second method was favoured, even though it yielded less compression, because it used about half as much memory.

Roberts found that the LOEMA algorithm could reliably identify changes in the authorship of a document.

Roberts's work addressed many of the fundamental issues in the field of Markov modelling. However, the LOEMA algorithm requires arbitrary blending between many orders of model, making it inefficient. The PPM algorithm (Section 1.10.6.6) uses a simplified form of blending and yields comparable compression.

1.10.6.5 UMC

One of the most impressive theoretical aspects of the LZ78 algorithm was its capability to converge on any ergodic source at infinity (universality). In 1983, Rissanen used the same underlying mechanism (of infinite tree growth) to construct a universal *Markov* algorithm which we will call UMC[46]. Because Markov algorithms are more powerful that dictionary algorithms, Rissanen's algorithm converges on the source much faster than LZ78.

The UMC algorithm compresses a binary (i.e. $n = 2$) source. The algorithm builds a backwards tree that reflects the properties of the source and then predicts from it. The tree is built as follows.

Each node in the tree has an associated sample. The algorithm starts with just the root node, whose sample is $(1, 1)$. The tree is uniformly 2-furcated. The tree is updated by adding the new instance to the sample of each matching node that already contains an instance of the new symbol. Call the deepest such node T. If T is a leaf, n child nodes are appended to T. Regardless, the child nodes of T are then incremented.

The algorithm takes an unusual approach to memory management and prediction. Tree growth stops when the deepest node does not reduce the tree entropy by a certain amount. At each step, the deepest node that *does* provide a sufficient entropy loss is used to make the model's prediction.

The algorithm can be seen to be universal by observing that at infinity, the count on a node differs from the number of times that it has occurred by an amount related only to the string associated with the node, not to the length of the history.

An interesting feature of the algorithm is that it accesses the history through a permutation mapping. This means that the algorithm can easily be configured to place different emphasis on instances at various distances.. This makes it easy to modify for the compression of two dimensional data such as images in which a pixel i will exhibit greater correlation with pixel $i - r$ (where r is the width of the image) than with pixel $i - 2$.

The algorithm is also unusual because it does not necessarily update to the greatest depth that it can. In fact the algorithm can be considered to be managing two tree structures, one for each symbol.

---

[46] In [**Rissanen83**] the only name given to the algorithm is "context" (p. 659). The author of this thesis (in his temporary capacity as algorithm trade mark registrar) thinks that the word "context" is too generic. The acronym "UMC" (for Universal Markov Compression) will be used instead.

1.10.6.6 PPM

The PPM (for Prediction by Partial (string) Matching) algorithm by Cleary and Witten[Cleary84] was the first practical variable-order, finite-context Markov algorithm. The PPM algorithm maintains $m+1$ Markov models of orders $Z[0, m]$. Samples are created only for contexts whose string appears in the history. These contexts are stored in a forwards tree structure.

Cleary and Witten do not directly address the problem of memory running out except to note that "... empty store is becoming a cheap resource. The major expense associated with memory is the cost of filling it with information and maintaining and updating that information." ([Cleary84], p. 401) When dealing with this algorithm, the tradition[47] seems to be to assume that infinite memory is available.

The defining aspects of PPM are its maintenance of models of different order and its blending of the predictions of the various order models. PPM differs from most other Markov techniques because it does not base its predictions on the sample of a single context. Rather, it calculates its prediction by blending samples from the different order models that it maintains. PPM uses a blending technique that is more time efficient than LOEMA's.

The PPM algorithm starts with a probability of 1 and allocates portions of the remaining probability until all symbols have been given a positive probability. Control starts with the highest order (m) model and works down to an order -1 model.[48] At each stage, a portion of the remaining probability is allocated to each symbol that has a positive frequency in the current order model and which has not already been allocated a probability.

The blending method is most concisely described by a program fragment (**Figure 15**). The function accepts an array of samples xs (one sample for the matching context at each depth) and returns an array of probabilities. During execution, a *pred* value of *notdone* indicates that a symbol's probability has not yet been calculated. The array xs contains an element indexed by -1 that has the value $xs(-1)(a) = 1/n$, $\forall a \in A$. This "uniform prediction" is used to "catch" symbols that have a zero

---

[47] As ascertained from private communication with Witten, Bell and Moffat.

[48] An order -1 model never records any instances and always makes the uniform prediction $p(a) = 1/n$, $\forall a \in A$.

```
function PPMest (in xs : samples) return prediction;
    pavail : real ← 1.0;
    notdone : constant real ← ∞;
    pred : array(symbol) of real ← (others → notdone);
begin PPMest
    for order in reverse −1 ... m loop
        total : integer ← 0;
        x : sample ← xs(order);
        for a in A such that pred[a]=notdone and x(a)>0 loop
            total←total+x(a);
        endfor;
        for a in A such that pred[a]=notdone and x(a)>0 loop
            pred[a]←pavail×(x(a)/(total+1));
        end for;
        pavail←pavail×1/(total+1);
    end for;
    return pred;
end PPMest
```

The PPM estimation algorithm works from the tip of the matching branch to the root, allocating the available probability as it goes. At each stage, the algorithm uses nonlinear estimation with $\lambda = 1$ to divide the remaining probability among the groups of zero and non-zero frequency symbols. At each level, the probability allocated for symbols with a zero frequency is called the *escape probability*. Implicitly there is an order -1 model whose only context contains a single instance of each symbol.

Figure 15: The PPM estimation algorithm.

frequency at all higher orders (i.e. symbols that have never appeared). The prediction result appears in *pred*.

This description was constructed for the purposes of exposition. Much more efficient implementations exist. In general, Markov algorithms are considered to require more resources than other classes of data compression algorithm. The PPM algorithm appears particularly inefficient because of the way in which it blends more than one sample together to form predictions.

Moffat[Moffat88] evaluated methods for improving the performance of the PPM algorithm. Successful modifications were: maintaining pointers to shorter matching contexts, using a specially optimized estimation scanner for the deepest context (which involves no exclusion), using a move-to-front list as a representation for predictions, maintaining a bit array to keep track of exclusions, count scaling, only updating the deepest matching node, reconstruction when memory runs out, and dispensing with exclusions. The final program used 512K of memory and processed about 4K of data per second without much loss of compression (in comparison to slower implementations). A highly tuned Ziv and Lempel algorithm on the same machine ran eight times faster but gave poorer compression.

1.10.6.7 DHPC

The DHPC algorithm is an algorithm developed by the author of this thesis[**Williams88**]. DHPC was developed independently from all the other algorithms and has advantages and disadvantages over the other algorithms.

DHPC is the same as PPM except for a few major changes. Whereas PPM blends predictions from the whole matching branch, DHPC bases its prediction upon the sample of the deepest matching node whose sample contains more than a constant threshold number of instances. Whereas PPM grows each branch of its tree to the maximum depth whenever a new context appears, DHPC grows its tree slowly using threshold counts to retard growth. PPM and DHPC also differ in the method of implementation adopted by their designers. PPM uses a forwards tree. DHPC uses a backwards tree.

Further discussion of DHPC is deferred until Chapter 2.

1.10.6.8 DMC

The DMC algorithm[**Cormack87**] represents a totally new approach to the Markov data compression problem. Instead of explicitly manipulating trees whose nodes correspond to finite context strings, the DMC algorithm manipulates a finite state machine *whose nodes are not particularly associated with anything!* Although the algorithm works with a symbol set of any size, it is strongly oriented towards the binary alphabet and will be presented here in binary form.

The DMC finite state machine consists of one or more nodes connected by arcs labelled 0 or 1. It is an invariant of the machine that each node has exactly two outward arcs, one labelled 0 and the other labelled 1. Associated with each arc labelled a is a transition count n_a which is the number of times that the arc has been traversed. At any point of time (i.e. in between instances), there is a distinguished node called the "current node".

The DMC algorithm starts with a single node that points to itself.
Each arc is assigned a transition count of one.

Figure 16: DMC starting configuration.

(Redrawn from figure 3(a) of [Cormack87])

The algorithm starts off with a machine consisting of a single node whose outward arcs point to itself (**Figure 16**). Each arc begins with a transition count of one. The single node is the current node. Before each instance is read, a prediction of the instance is generated based upon the transition counts of the outward arcs of the current node. The estimation formula is $p(a) = (n_a + c)/(n_0 + n_1 + 2c)$ where c is the usual smoothing constant.[49] This is linear estimation with $\lambda = cn$ $(n = 2)$. After the instance (a) is read, the current node changes to the node at the end of the arc labelled a.

Thus far, the algorithm is fairly standard (as finite-state machines go). The unusual aspect of DMC is the way its grows its finite state machine. Each time that a transition is about to be made, a check is performed to see if the transition count of the arc about to be traversed exceeds a certain constant threshold t_1. If it does and the transition count of all other inward arcs to the next node exceeds a similar constant threshold t_2, then a **cloning** operation takes place. Otherwise it does not.

The effect of the cloning operation is described by the following five pointer assignments which must be executed sequentially. The **newnode** function creates a new node and returns a pointer to it. *newn* is a temporary pointer that is used to point to the new node. *curr* is a pointer to the current node just before the cloning takes place. *next* is a pointer to the node that would be the next current node if the cloning did not take place. The dot notation is used to refer to the two pointer fields (named 0 and 1) of a node.

$$newn \leftarrow \mathbf{newnode};$$
$$next \leftarrow curr.sym;$$
$$curr.sym \leftarrow newn;$$
$$newn.0 \leftarrow next.0;$$
$$newn.1 \leftarrow next.1;$$

---

[49] Cormack and Horspool do not specify the value of c that they used in their experiments.

As well as adding a new node and connecting the node into the machine (as described above), the cloning operation apportions some of the counts of the outward arcs of the next node to the outward arcs of the new node. An example of the cloning operation (not giving counts) is shown graphically in **Figure 17**.

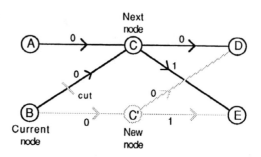

Cloning occurs in the DMC algorithm just before transitions are about to be made. In this case the transition was to be from B to C. However, as the transition counts on arcs AC and BC exceeded specified thresholds, a cloning operation occurred. A new node C' was created and arc B0 switched to C'. The effect was to divide the context of C between C and C', making the contexts more specific.

Figure 17: DMC cloning operation.

(Based on figure 2 of [Cormack87])

Although the cloning operation appears simple, its effect can be quite complex because of aliasing. In **Figure 17**, all the nodes are distinct. If nodes are aliased, the operation is qualitatively different. In particular, if *curr* and *next* turn out to be the same node, the newly created node winds up pointing to itself (**Figure 18**). This is a consequence of the third assignment being executed before the fourth and fifth assignments (in which the new node copies over the output pointers of the current node).

To the casual observer it may seem that this algorithm does not make Assumption 7 (Section 1.10.2) and that it is capable of employing arbitrary state machine structures to model the source. This conjecture has been proven incorrect by Bell and Moffat[Bell88] who showed that the only finite state automata that DMC is capable of generating are also finite context automata. This means that a finite context string can be associated with each node. Experimental results confirm that DMC has about the same power as the other Markov algorithms[Bell89].

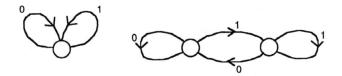

Cloning is not so simple if the current node points to itself. This case arises at the start of the algorithm (shown here) where the first node is always cloned into two new nodes (with context strings 0 and 1) that point to themselves and each other.

Figure 18: Cloning a node that points to itself.

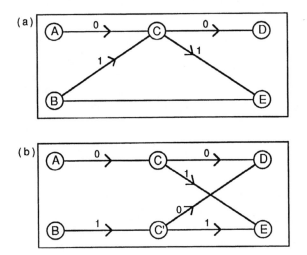

So little is known about the DMC algorithm that the main tutorial diagram of the original paper is at best misleading and at worst incorrect. In DMC, a node can have heterogeneously labelled inward arcs only if it is the original node and no cloning has taken place. But in this diagram, node C has heterogeneously labelled arcs.

Figure 19: Erroneous tutorial example of DMC cloning.
(Redrawn from figure 2 of [Cormack87])

Not much else is known about the behaviour of DMC. In fact, so little is known that it turns out that the main tutorial diagram that the inventors of the algorithm use to describe the cloning operation is at best misleading and at worst incorrect. Figure 2 in section 4.3 of [Cormack87] (redrawn here in **Figure 19**) contains two subwindows ((a) and (b)) which are supposed to illustrate a typical cloning operation. In (a) the node that is about to be cloned has two inward arcs labelled 0 and 1. In (b) the node has been cloned. In contrast, the DMC algorithm ensures that the only time that a node can have heterogeneously labelled inward arcs is if the node is the initial node and then only if it has not yet been cloned. This means that all five nodes in (a) must be the same node. This makes nonsense out of (b). The figure would make sense if both inward arcs in (a) were labelled 0 (as in **Figure 17**). Bell and Moffat use the same example in their commentary paper[Bell88] but appear not to have made the same mistake.

Cormack and Horspool could well be forgiven for this error. Once the algorithm starts running, it is not at all clear what actually happens. DMC is one of those objects whose workings are extremely simple but whose results are extremely complicated.[50]

1.10.6.9 WORD

The Markov techniques described so far all parse the input into instances. Although parsing into instances is the best approach in theory (Section 1.8.3), the practical advantages (such as increased execution speed) of treating the input at a higher level justify further investigation.

Moffat[**Moffat89**] investigated the application of Markov models (such as PPM) at the word level. A "word" is defined to be a maximal sequence of alphabetic characters. A message is considered to be a sequence of alternating words and non-words. Moffat's WORD algorithm parses the message into a stream of words and a stream of non-words and compresses them separately at the word level using two separate Markov models. Markov models are also maintained for the length and character-level characteristics of words and non-words. This allows new words to be compressed as they are spelt out.

WORD works well for both text and non-text data. For non-text data, the scheme degrades gracefully to a character-level Markov model. Moffat experimented with various depths and found that order-one models out-performed order-zero models (by about 13%) but that order-two models did not yield a significant advantage over order-one models.

WORD is one of the best Markov algorithms. Its performance is comparable to DMC and PPM but it runs faster.

This section ends the review of Markov algorithms. Attention now turns to their implementation.

---

[50] Other examples are Rubik's cube[**Singmaster80**] and Mandelbrot sets[**Barnsley88**].

1.11 Data Structures For Predictions

Predictions are the glue that connects models to coders. Whenever a new instance is about to arrive, the model generates a prediction and hands it to the coder (**Figure 9**). The coder then uses the prediction to code the incoming instance. The prediction contains a probability for each symbol.

A major problem with handling predictions is their size. Predictions for the binary alphabet consist of a single probability and can be manipulated very efficiently. For larger symbol sets, predictions consist of a vector of probabilities which can be quite large. For $n = 256$ and probabilities of just one byte, predictions will consume 256 bytes. This makes them impractical to create, copy (e.g. from model to coder) or destroy on a per-instance basis.

Given that it is forbidden to construct a new prediction for each instance, the stream of predictions being passed to the coder must take the form of *views* to another data structure *already created*.

The obvious candidate for the "already created" predictions are the samples stored in the nodes of the Markov models (x in Section 1.8.2 and Section 1.10.2). Our present description of x as an array of symbol frequencies is totally oriented towards the Markov model. By re-organizing the data structure so as to serve the coder's needs, the structure can be used to connect the model to the coder without ever copying predictions.

1.11.1 Prediction Functionality

Before looking for a suitable data structure for predictions, it is worth specifying exactly what a prediction data structure must do. This section describes a prediction abstract data type[**Guttag80**] that serves the needs of both model and coder.

The Markov model's requirements of the prediction are very simple. All that the prediction must do for the model is to maintain a frequency for each symbol. Some specialized models might require more operations but the basic functionality is described as follows. p denotes the prediction object.

init(p) — Initializes the prediction's sample, giving every symbol a frequency count of zero.

inc(p,a) — Increments the frequency associated with symbol a.

In contrast, the arithmetic coder's requirements are quite demanding. In order to divide the code space (Section 1.9.1) among the symbols, the coder has to define an ordering[51] on the symbols. The actual ordering used by the coder does not matter as long as the decoder can reconstruct the ordering. This ordering is used to determine the symbols above or below a given symbol when dividing up the code space. To code an instance of a particular symbol, the coder must find the sum of the probabilities of the symbols above the target symbol. In practice, the probabilities are manipulated in the form of frequency counts. The requirements of the coder and decoder are specified as follows.

symbol_to_range(p,r,a) — Returns a tuple being the first and last element in the subrange allocated to symbol a, from the range $[1, r]$.

range_to_symbol(p,r,s) — Returns the symbol corresponding to the subrange that contains integer s, from the range $[1, r]$.

Thus, one end of the prediction data structure absorbs instances and the other end generates ranges. Hidden somewhere in the middle is an estimation technique (Section 1.10.3) which ensures that symbols of frequency zero are never mapped into an empty range. With a few minor modifications, most of the data structures to be described can be modified to use any estimation technique.

The requirements above define the prediction data structure. The major difficulty with finding an efficient representation is the conflict between the model's need to modify the count of random[52] symbols (*inc*) and the coder's need to access the *sum* of frequencies of symbols above a random symbol (*symbol_to_range*, *range_to_symbol*). Unfortunately, any explicitly stored information about the sum of frequencies above a given symbol can be invalidated by a single *inc* update to a higher symbol (**Figure 20**). We cannot expect a constant time solution.

---

[51] It is usual to use the words "before" and "after" to refer the relation between different symbols (e.g. "symbol x occurs *before* symbol y in the set of symbols"). Unfortunately the temporal flavour of this nomenclature invites confusion between the ordering of the set of symbols and the ordering of instances in the message. Accordingly, we will use "above" and "below" ("higher" and "lower") to refer to the ordering of symbols within the set of symbols and use "before" and "after" ("previous" and "subsequent") to refer to the ordering of instances in the message.

[52] "Random" here is used in the sense of *Random* Access Memory; that is, unpredictable and arbitrarily chosen accesses.

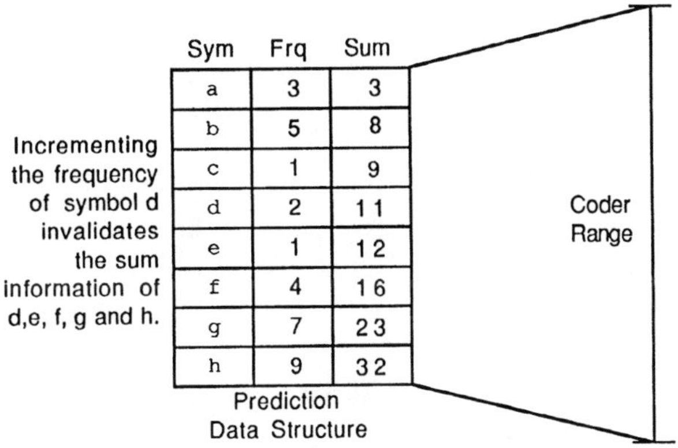

Sym	Frq	Sum
a	3	3
b	5	8
c	1	9
d	2	11
e	1	12
f	4	16
g	7	23
h	9	32

Incrementing the frequency of symbol d invalidates the sum information of d,e, f, g and h.

Coder Range

Prediction Data Structure

The conflicting requirements of models and coders mean that an array cannot be used to store frequencies with temporal efficiency. Because the coder requires the sum of frequencies above a particular symbol (so that it can map symbols to ranges), incrementing the frequency of a single symbol requires an $O(n)$ update of the sum array.

Figure 20: Problem with maintaining frequency sum information.

1.11.2 Linear Representations

We start by considering linear data structures (such as an array (**Figure 20**) or a linked list) for which all operations take time linear in the number of symbols. Although such linear structures are undesirable in a design in which prediction copying is forbidden, two significant improvements make the data structure feasible. Both improvements apply to more sophisticated data structures as well.

• Storage space and search time can be improved by using a *sparse* data structure and storing only those symbols whose frequency is positive. In practice, most samples are sparse.

• Search time can be improved by *ordering* the list so that the most frequently used symbols are near the front. Proposed list management heuristics are **frequency order, climb** and **move-to-front**.[53]

---

[53] In *frequency order* management, the list is maintained in frequency order. In *climb* management, the record for a symbol is moved one position towards the head of the list whenever the record is accessed. In *move-to-front* management, the record for a symbol is moved to the head of the list whenever the record is accessed.

It is unclear which of the alternatives of the second technique should be used. However, it should be noted that the climb and move-to-front heuristics can be performed in constant time whereas frequency ordering can degenerate to time linear (or if an array and binary searching are used, logarithmic) in the number of symbols.

Linear data structures (with the given improvements) work well for sparse, highly skewed distributions. However, data compressors should be designed to handle white noise gracefully and so it is desirable to look for data structures with a better worst-case running time.

1.11.3 Sparse Representations

Before turning to tree representations, it is worth investigating the issue of sparseness in greater detail.

One of the axioms of Markov algorithms is that, in general, a higher order model can model a source better than a lower order model can. This means that for a redundant source, we can expect that on average the entropy of deeper nodes of the Markov tree will be lower than that of shallower nodes. Low entropy distributions are characterized by "spikiness". This in turn manifests itself in *sparsity*. As an example, in English text, it is common for the samples of deep nodes to contain only 10 symbols with a positive frequency count. For 256 symbols, this represents a 95% sparsity.

The greatest danger of using sparse data structures is the lack of any guarantee of sparsity.[54] Because sparse data structures use up more memory per element than non-sparse data structures, sparse data structures are likely to use *more* memory than a non-sparse data structure for non-sparse data. To choose a sparse representation is to gamble memory on the entropy of the source.

Consider a Markov tree algorithm that has just run out of memory. If a fixed size representation for samples is being used, the algorithm is guaranteed not to require any more memory. It can (say) freeze its tree and continue to record instances. On the other hand, if a sparse representation is being used, more memory may be needed. The only options are then to destroy part of the Markov tree, or to recycle the records of infrequent symbols in some samples.

One technique is to switch between sparse and non-sparse data structures in accordance with the data. This does not solve the problem

---

[54] The variable size of coded data is a problem with data compression in general.

of what to do when memory runs out but it does eliminate the spatial disadvantages of sparse representations, admittedly at the cost of extra processing time.

In practice, data is sparse, and sparse data structures are nearly always worthwhile.

1.11.4 Tree Representations

Tree structures provide a better solution to the prediction representation problem, generally solving it in time logarithmic in the number of symbols. Each symbol is stored in a single node of a binary search tree. Each node is tagged by a symbol and stores *the sum of the frequencies of all the symbols in the node's subtree* (including the node itself) (**Figure 21**). For a leaf node, this sum is the same as its symbol's frequency. For a non-leaf node it is the sum of its symbol's frequency and the sums of its child nodes. The *inc* operation is performed by traversing from the root to the node of the target symbol, incrementing the sum of each node visited. The *symbol_to_range* and *range_to_symbol* operations traverse the tree from the root using the symbol and sum values to extract symbols or ranges.

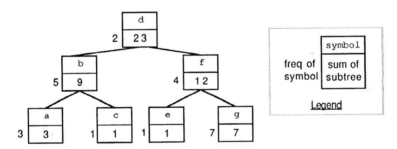

A tree structure representation of samples/predictions satisfies the needs of the model and coder in $O(\log n)$ time. Leaf nodes store the frequency of their symbol and non-leaf nodes store the sum of their frequency and the numbers in their child nodes. Unlike the array implementation, incrementing a symbol affects only the symbol's node and its ancestors. The coder can obtain ranges by moving down the tree.

Figure 21: Prediction tree structure containing subtree sums.

The two optimizations used to improve linear representations can be profitably applied to tree structures. Sparse trees are easily implemented and provide considerable space savings. The ordering of a list structure corresponds to minimizing the tree path of the nodes of frequently used symbols. Tree balancing is an involved field in its own right and we will only review the most relevant tree management algorithms.

The simplest tree management is no management; that is nodes are never moved after they are attached to the tree. The disadvantage here is that a badly built tree can yield linear performance — as poor as a list. However, so long as the worst case performance is not fatal to the compression system, the worst case may not be much of a threat in practice. For an ordinary source, on average, the earliness of the first occurrence of an instance of a symbol will be proportional to the symbol's probability. This means that it is unlikely that a degenerate tree will be built. A simple one-to-one hash function can be employed to avoid the worst case occurring for common orderings (such as sorted sequences).

The next class of tree management dynamically manipulates the tree so as to optimize characteristics such as the tree's depth. The problem of balancing trees is old, and many structures exist including height balanced trees, weight balanced trees, b-trees, optimal search trees, biased search trees and finger search trees[Sleator85].[55] All of these trees have disadvantages. We do not intend to explore this involved field. Only some of the solutions will be reviewed.

Of the balancing techniques, one of the simplest is height balancing (AVL balancing)[Wirth76](p. 215) which endeavours to reduce the height of the nodes in the tree. Height balancing guarantees logarithmic performance but is sub-optimal for skewed distributions.

It might appear that for skewed distributions dynamic Huffman trees (Section 1.4.4) would minimize the average path length. Unfortunately, because Huffman codes possess the prefix property, dynamic Huffman trees do not utilize internal nodes and their search path is about one arc longer than that of other trees.

The recently developed splay trees by Sleator and Tarjan[Sleator85] provide an excellent prediction representation. Splay trees are self adjusting binary search trees that have the following property.

"For an arbitrary but sufficiently long sequence of retrievals, a splay tree is as efficient to within a constant factor as an optimum static binary search

---

[55] See [Tarjan83] for a review of the current state of tree and graph algorithms.

tree expressly constructed to minimize the total retrieval time for the given sequence." **[Tarjan87]**(p. 211)

Splay trees enjoy the following advantages over other trees.

- Splay trees are simple.

- Splay trees do not require storage for balancing information.

- Splay trees adapt very quickly to changing access characteristics.

A recent paper by Jones[Jones88] investigated applications of splay trees to data compression. The first part of the paper describes an optimized splay tree algorithm which can be used to construct an efficient adaptive prefix code similar to dynamic Huffman coding. The optimizations are possible because with a few additional data structures, the tree need not be ordered lexicographically. The second part of the paper investigates the use of splay trees for representing predictions. Jones formalizes the prediction representation problem in much the same way as we have here, and uses the basic cumulative frequency tree structure described above. The differences are as follows.

- Only the leaves of the tree are used to represent symbols.

- The cumulative frequencies are used for the coding but have no effect on the structure of the tree. The splaying algorithm is used to balance the tree.

- The modified splaying algorithm is used.

The result is a fast algorithm that provides an *amortized*[Tarjan85] time complexity that is *linear* in the length of the code string. This is a fascinating result because the length of the code string and the processing time are determined by independent processes (the cumulative frequencies and the splaying) driven by the same data.

It is unclear why Jones did not utilize the internal nodes in the tree. This omission may simply have been a carry over from the prefix trees described in the first part of the paper.

The splay tree solution yields a linear *amortized* bound. This means that at any instant, the number of output nats may not correspond to the time taken to produce them; in practice a buffer would be required to drive a fixed rate channel.

1.11.5   Heap Representations

Moffat[Moffat88] has provided a prediction data structure that yields the same linear bound as splay trees, but at the instance level (i.e. not amortized).

The scheme uses four arrays of equal length, the length being the number of symbols.   The first array records the frequency of each symbol in frequency order.   The next two arrays provide mappings between symbols and their corresponding elements of the frequency array. Because the first array is in frequency order, it can be considered to be a perfectly balanced heap. The fourth array stores the sums of frequencies of each element's subtree in the heap. The *inc* operation is performed by swapping the target symbol with the leftmost (high frequency direction of the frequency array) symbol with the same frequency (logarithmic time using binary search). The frequency of the target symbol is incremented (constant time) and the cumulative frequency branch of the heap is incremented (logarithmic time).   This scheme operates in a manner similar to the Dynamic Huffman Coding algorithm (Section 1.4.4).

In this algorithm the access time and the length of the output string are controlled directly by the same mechanism — the frequency array. This yields an instantaneous linear time complexity in the length of the output, removing the need for an output buffer. This is the only prediction data structure to date with this property.

One of the problems of using large symbol sets is that wide coder ranges are required for accurate coding. For example, if there are 256 symbols, each with a probability that is an integer multiple of $1/1000$, the coder range would have to be at least $256000 \times k$ wide to guarantee an accuracy of one part in k within each probability division ($1/1000$). Such large ranges stress coder designs.[56]

An alternative to coding large-alphabet symbols directly is to use a binary arithmetic code (Section 1.9.4) to transmit binary branching decisions on the path from the root to the target leaf, using the cumulative frequencies of each pair of candidate child nodes as probabilities. This technique is called decomposition[57] and has been proposed as a technique for coding predictions[Langdon83]. Decomposition allows the use of alphabets of unbounded size without introducing precision problems.

---

[56] For example, the author's own coder implementation uses double-precision floating point numbers.

[57] The technique of "decomposition" comes from [Shannon48](section 6, figure 6) who noted that his definition of information allowed a decision tree of arbitrary furcation to be converted into a binary decision tree with the same entropy.

1.11.6 Stochastic Representations

All the prediction representations discussed so far are oriented towards increasing the speed of the data structure. In general, speed is of greatest importance in a prediction data structure. However, because Markov models can have many thousands of contexts, memory consumption is also important.

A large portion of the memory consumed by a prediction is taken up by frequency counters.[58] The fact that predictions are based only on the *relative* frequency of each symbol in a sample suggests that a floating point representation could be used to reduce the memory consumption of these counters.

An 8-bit counter of the form $m \times 2^e$ with $m \in \mathbf{Z}[0, 15]$ and $e \in \mathbf{Z}[0, 15]$ can store natural numbers (scattered exponentially) in the range $\mathbf{Z}[0, 15 \times 2^{15}]$. This is more than enough to prevent overflow under normal conditions. The major problem is incrementing a counter with a positive exponent.

One possible implementation of the increment operation is to increment the mantissa m with probability 2^{-e}. This organization conserves memory at the expense of accuracy and uncertainty in the value of the count; after a number of increment operations, the counter's value will be spread in a binomial distribution around the ideal value (i.e. the value the counter would have if it was an integer counter of infinite width).

The technique of stochastic counting presented above was devised by the author of this thesis in early 1987. A less general form of the same idea was devised in 1982 by Helman, Langdon, Martin and Todd [**Helman82**] who proposed stochastic incrementing as a method for updating skew numbers (Section 1.9.4). A skew number is essentially a binary exponent (e in the above). Despite the lack of a mantissa, the scheme is efficient for binary codes because of the accuracy to which a binary prediction can be approximated by a power of two (Section 1.9.4). The technique of stochastically incrementing skew factors was successfully incorporated into the adaptive Q-Coder described in Section 1.11.7.

It should be mentioned that decoding can still take place in a compression system that utilizes random numbers so long as a *deterministic* random number generator is used. Linear congruential generators are the most popular and are described in [**Knuth81**] (chapter 3). Recently

---

[58] The pointers of sparse representations also use up a lot of memory.

a minimum standard for random number generators has been proposed [**Park88**]. In most compression systems, speed is likely to be more important than randomness and faster, less sophisticated techniques may be more effective.

If storage is really at a premium, samples could be represented as a list of the frequencies of each symbol with each frequency f stored as $b = \lceil \log f \rceil$ 0s followed by a 1 followed by the b bit representation of f. A sample of y instances would require a minimum and maximum of

$$\mathbf{Z}\left[n + 2\lceil \log_2 y \rceil, n\left(1 + 2\lceil \log_2 y/n \rceil\right)\right]$$

bits depending on the entropy of the sample. By sharing space between counters, this representation requires space logarithmic in the total number of instances. However, as prediction data structures are already speed-stressed, this structure is unlikely to be practical in the immediate future.

1.11.7 The Adaptive Q-Coder

One approach to prediction representations is to roll the coder, prediction and context together into a single data abstraction that receives instances and transmits channel symbols. Traces of this approach are present in Moffat's presentation of his prediction data structure [Moffat88], although strictly Moffat's structure conforms to the prediction data structure specification given earlier (Section 1.11.1).

The adaptive Q-coder[Langdon88][Mitchell87] combines several ideas together to form a powerful integrated binary arithmetic coder. At the heart of the method is an optimized version of the simple binary arithmetic code of [Langdon82] (Section 1.9.4). The new code is very fast, involving only shift and addition operations.

Driving the arithmetic code is a stochastically incremented sample that consumes only six bits. One bit stores the most probable symbol (0 or 1). The other five bits store a skew number.[59] The complete six bits form a prediction. Whenever the least probable symbol arrives, the skew number is decremented and the coder register is shifted. Whenever the most probable symbol arrives, a value determined by the skew number is added to the coder register. If the coder register has to be shifted, the skew number is incremented. The effect is that the coder register acts as mantissa to the skew value's exponent.

---

[59] In the Q-coder, the relationship between the skew number and the coder is more complicated than a simple power of two. Here, the description is simplified for the purposes of exposition. For full details of the Q-coder refer to [Langdon88].

The symbiosis of the Q-coder's skew number and coder register suggests that the Q-coder could not be modified to operate with more than one context. This is not so. Although a single-context Q-coder apparently relies on the coder register as a mantissa, it can do nearly as well with a stream of random numbers. This enables many contexts to share the same coder register. The result is a single-coder, multi-context system requiring only six bits per context.

1.11.8 A Comparison of Representations

The multitude of prediction representations begs a comparison. Unfortunately, nobody has has performed a comparative study. Apart from the theoretical results, the only practical result seems to be that sophisticated data structures usually do not perform as well as the simple ones [Moffat89](p. 191). In practice alphabets are sparse, and worst cases rarely occur. At the end of his paper on splay trees, Jones[Jones88] compared his splay tree structure with a move-to-front list structure[60] and found that the list ran faster for entropies less than 6.5 bits/instance. 6.5 bits is a very high entropy for ordinary byte data and so it seems that simple data structures run faster in practice. Moffat came to much the same conclusion[Moffat89](p. 191). The author's own experience with splay trees supports this result.

1.11.9 Summary

Predictions form the glue that connects the model and the coder. This section has established prediction data structures as an important component of modern data compression implementations. Prediction data structures are subject to tight specifications which make them difficult to implement efficiently (i.e. in constant time). Because prediction abstractions map instances to ranges, they must embody an estimation technique. Many representations can be used for predictions, ranging from ordered lists and balanced trees to the sophisticated splay and heap data structures recently proposed. For practical data compression using small alphabets, sophisticated data structures may be less effective than simple data structures. Tree structures allow binary decomposition which can be used to avoid high precision arithmetic.

---

[60] Jones claims to have used the "move-to-front" C program given in [Witten87]. However the text (p. 536) and program (p. 531) in [Witten87] describe a frequency ordered list. As a consequence, it is unclear whether Jones was using a move-to-front or a frequency-ordered heuristic.

1.12 Data Structures for Markov Trees

The previous section discussed the sample/prediction data structure which is associated with each context in a Markov model. This section zooms out to cover representations for the Markov tree structure itself.

The constraints on representations for Markov tree structures are much less stringent than those for the representation for predictions. The tree structure is fundamentally a one-way mapping from strings to nodes. There seems to be two ways of representing it. The first is hashing, in which strings are mapped directly to nodes. The second approach is to use an explicit tree structure. Both forward and backward trees can be used.

1.12.1 Hashing Representations

A very simple method for storing the nodes of a tree is simply to hash string values directly onto their corresponding nodes. This method was used in the LOEMA algorithm[Roberts82](chapter 4, p. 32). To use a hash table, a hashing algorithm must be found that maps strings of length $\mathbf{Z}[0, m]$ onto numbers in the range $\mathbf{Z}[0, h-1]$ where h is the length of the hash table. Each hash table entry corresponds to a Markov tree node. Hash tables suffer from the disadvantages of being of a fixed size[61] and of having to store the key value in each entry. In this case, strings of up to length m must be stored. Although this may seem a problem, in practice it is not. Experiments show that the optimal m is about 3 (Experiment 5 in Section 4.17.8, [Moffat88]). For $n = 256$ and $m = 3$, a maximum of three bytes would be needed to store each entry.

Hash tables have the advantage of allowing totally random access to any node in the tree. So long as the maximum node depth is kept low, hash tables will conserve memory by eliminating pointers. An advantage of hash tables is that they open up the possibility of storing non-tree structures (e.g. omitting parent nodes).

1.12.2 Backwards and Forwards Trees

The alternative to using a hash table is to construct an explicit tree structure (either in an array or in heap). Tree structures can be constructed incrementally (unlike hash tables which must be completely initialized at the start of the run) and provide fast (pointer) access between related (parent/child) nodes.

---

[61] However, extendable hash tables have recently been devised.

The main design decision in representing a tree structure with a tree structure is the way in which child nodes are connected to their parent node. Most of the discussion in Section 1.11 applies here as well. The main difference between **furcation trees** and prediction trees is that furcation trees need not conform to the requirements of the arithmetic coder. Further discussion of furcation implementations can be found in Section 4.6.

The next major design decision is whether to use forward trees or backwards trees (Section 1.2, **Figure 3** (reproduced in **Figure 22**)). As each kind of tree is capable of mapping strings to nodes, each is capable of implementing the required mapping. The author's DHPC algorithm (Chapter 2) is strongly based on backwards trees. In contrast, all implementations of PPM and its variants have used forwards trees ([Cleary84][62], [Bell89], [Moffat88]).

The choice between backwards and forwards trees is considerably muddied by proposed improvements involving cross-tree pointers. We start by considering trees (**Figure 22**) without such pointers.

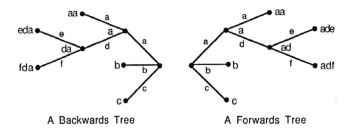

A Backwards Tree   A Forwards Tree

In backwards and forwards digital search trees each node corresponds to the string constructed by moving from the root to the node. The root node corresponds to the empty string. The only difference between backwards and forwards trees is the direction of their strings. This figure is a duplicate of **Figure 3**.

Figure 22: Backwards and forwards digital search trees.

---

[62] This first paper did not give many details on the way PPM was implemented. However verbal descriptions by the authors along with a paper by Moffat[Moffat88] and a book by the inventors[Bell89] of the algorithm confirm their use of explicitly linked forward trees.

The basic requirement of a tree is that it provide fast access to all nodes whose strings match the history. In the case of a backwards tree, the set of matching nodes always form a branch leading out from the root. If the maximum depth of the tree is m, all matching nodes can be located using m arc traversal operations (hops). In the case of a forwards tree, the matching nodes are scattered through the tree and must be looked up separately. For a history string "sloth", the nodes "h", "th", "oth", "loth" and "sloth" must all be accessed separately. This takes $m(m+1)/2$ hops.

Bell, Cleary and Witten[Bell89] proposed a technique for reducing the $m(m+1)/2$ hops. A window buffer of $m+1$ pointers is maintained that points to the currently matching nodes of depth $Z[0, m]$. Whenever an instance arrives, each pointer is moved one arc down the tree and the entire buffer is shifted. The pointer of depth m drops out one end of the buffer and a pointer to the root node is inserted at the other. This system cuts the per-instance branch access time down to m hops, which is the same as for backwards trees.

So far backwards trees appear far superior. Backwards trees provide access to the matching branch of any random history in m hops. Forwards trees require $m(m+1)/2$ hops for a random history but for sequential data, yield m hops with the aid of an extra data structure. The main advantage of forward trees is that their furcation data structure can be combined with their prediction data structure to yield a memory saving for internal nodes.

In practice, once an algorithm has been running for a while, most predictions are made by the deepest node of the matching branch. This fact led to the independent discovery of **vine pointers** ([Bell89]) in forward trees and **shortcut pointers** (Section 4.10) in backwards trees. The vine pointer of a node as ($a \in A$, $s \in S$) in a forwards tree points to node s. Vine pointers eliminate the need for a window buffer of pointers. For backwards trees a shortcut pointer is associated with each symbol in each prediction. For symbol a of node s, the pointer points to node sa. To allow access to the entire matching branch, **parent pointers** must be stored as well.

At this stage, the backwards tree seems a little overloaded with data structures; in fact, for sequential data, the links from parent nodes to their child nodes are no longer required. If they are removed, the backwards tree turns into a forwards tree.

The relationship between backwards trees and forwards trees becomes clearer if each structure is viewed merely as a collection of nodes supporting one or more node-to-node mappings. Because each node corresponds to a context string, the mappings are best expressed as mappings between strings. **Table 3** lists four such mappings, their (logical) names and the names given to them by advocates of backwards and forwards trees. Mappings that require the specification of a symbol a are far more expensive ($O(\log n)$) than mappings that don't ($O(1)$). The entry $sa \longrightarrow s$ in the table is currently not used in any algorithm and is included only for completeness.

Mapping	Name	Backwards Name	Forwards Name
$s \xrightarrow{a} as$	*AddLeft*	Child pointer	—
$as \longrightarrow s$	*StripLeft*	Parent pointer	Vine pointer
$s \xrightarrow{a} sa$	*AddRight*	Shortcut pointer	Child pointer
$sa \longrightarrow s$	*StripRight*	—	Parent pointer

Table 3: Tree mappings and their names.

These mappings provide a simple tool for analysing tree structures. A basic backwards tree (as in [Williams88]) is described by {*AddLeft*}. A basic forwards tree (as in [Cleary84]) is described by {*AddRight*}. A backwards tree with parent pointers and shortcuts pointers (as in Chapter 4) is described by {*AddLeft, StripLeft, AddRight*}. A forwards tree with vine pointers (as in [Bell89]) is described by {*AddRight, StripLeft*}.

Different mappings lend different properties to a tree. The *AddLeft* mapping is needed if $O(m)$ random access to contexts is required. The *StripLeft* mapping is needed if context blending is to be performed. However, an array of pointers can be used to avoid it if the *AddRight* mapping is available. The *AddRight* mapping coincides with the prediction data structure. The *StripLeft* and *AddRight* mappings form a very powerful combination because they can very efficiently convert a history buffer string of the form as into the form sb ($s \in S, a \in A, b \in A$). The conclusion is that forwards trees are more efficient for flattish trees.

The distinction between backwards and forwards trees becomes less blurry when we consider the effect of maintaining *incomplete* trees. So far, we have assumed that a node is created for each distinct string of length m or less that appears in the history. Under such conditions, forwards and backwards trees store identical sets of strings and have an identical effect. If memory is restricted, leaf nodes must be destroyed (or never created). In such situations, forwards and backwards trees present different pruning alternatives.

In a forwards tree, the string `sloth` is stored with the h at the leaf end. In a backwards tree, the `s` is at the leaf end. In one case, removal of a leaf will leave the context `slot` and in the other `loth`. Similarly, when choosing where to place a node, for leaf node `lot`, a backwards tree will present `slot` and a forwards tree `loth`. In the end, it may turn out that both kinds of leaf addition operations yield equivalent compression on average. However, from an organizational standpoint, the options presented by the backwards tree are more attractive. In general, if a node is heavily used, it should be divided into nodes that represent more specific contexts. For a context whose string is s, contexts of the form as should be created; s defines the context and a refines it. In a backwards tree, specialization of a context can be performed by adding one or more child nodes to the heavily used node. In a forwards tree, an entirely new branch must be constructed to achieve the same effect. Whereas attaching a leaf node to a node s in a backward tree creates a context as, addition of a leaf node to a node s of a forwards tree creates a new context sa whose recorded instances will not be a subset of those of the parent context.

In conclusion, for flattish trees, forwards and backwards trees provide much the same performance. For sequential data, forwards trees emerge as the most efficient representation, mainly because they save memory by combining the furcation and prediction data structures. For randomly accessed contexts, backwards trees are the best.

For irregular trees, the emphasis moves from efficiency to organization. Backwards trees present the most coherent tree growing and pruning options because the tree structure corresponds directly to a refinement of contexts. The forwards tree structure does not link related contexts as closely.

In this thesis, backwards trees are used exclusively. Chapter 2 describes an algorithm that uses a backwards $\{AddLeft\}$ tree. Later on in Chapter 4, a more sophisticated algorithm that uses a backwards $\{AddLeft, StripLeft, AddRight\}$ tree. This is the most powerful structure available.

1.13 Dictionary Methods vs Markov Methods

The field of data compression is currently dominated by the two major classes of algorithm: dictionary techniques and statistical techniques. Dictionary techniques are represented by the adaptive Ziv and Lempel compressors, and in particular LZ78. Statistical techniques are represented by finite-context Markov models, and in particular PPM.

At present each class of algorithm is holding some ground. Markov algorithms yield the best compression but dictionary algorithms are much faster and are invariably chosen in practice. It is worth throwing some light on the relationship between these two classes of algorithm.

Langdon[**Langdon84**] developed the ideas presented in his earlier paper[Langdon83](Section 1.7.5) about the statistical equivalent of LZ78. The main result is that it is always possible to construct a Markov model that emulates a dictionary technique. The converse is not true.

Consider a dictionary of d strings each of which is assigned a probability p_i. For the sake of simplicity, we assume that a perfectly efficient arithmetic code is being used. The message is parsed by the dictionary using greedy parsing. If the dictionary is represented as a forwards ($\{AddRight\}$) tree, the parsing of a single phrase consists of traversing the tree from root to leaf, one arc per incoming instance. Associated with each node is a code string. By taking the inverse log of the code string lengths at the leaves and working backwards down the tree, probabilities can be assigned to each arc of the tree. Each forward branching then corresponds to a prediction. The context on which each "prediction" is based depends on the depth of the node in the tree. Because a parse consists of walking from the root to a leaf, successive "predictions" are made using successively deeper orders. When a leaf is reached, control returns to the root and the next prediction is a zero order one. The result is the generation of predictions with orders that vary as a saw-tooth function.

This is the fundamental difference between Markov techniques and dictionary techniques. Markov models apply the same amount of "power" to each instance whereas dictionary techniques apply power in a saw-tooth function. The effect is that Markov models yield better compression than dictionary techniques, but require more processing time.

Langdon quantified all these ideas in his paper by defining three models: M — a dictionary (parsing) model, $M0$ — the dictionary model expressed as a Markov model, and M' — a full Markov model using the same number of contexts as M. He then showed that M' will always perform at least as well as M, with equivalence occurring when the source structure actually does follow the saw-tooth order curve.

Langdon pointed out that Markov models can approximate a source as closely as desired as soon as the model order exceeds that of the source. In contrast, parsing techniques, which are based on the saw-tooth power function, must grow an infinite number of phrases in order to raise the saw-tooth period to the point where the inefficiency of the zero order predictions at the start of the saw-tooth have little impact. Langdon showed that for a universal parsing model, the length of phrases approaches infinity as the message length approaches infinity. This implies that the average order of the implicit Markov models approaches infinity as well.

Experimental results confirm Langdon's conclusion. **Figure 23** shows the performance of two LZ78 variants called LZC and LZFG and a PPM variant called PPMC.[63] LZC is the algorithm used in the Unix *compress* program. LZFG is the best LZ78 algorithm developed so far (in general LZ78 class algorithm perform worse than LZ77 class algorithms[64]). PPMC is the best Markov algorithm developed to date. The vertical axis plots compression (proportion remaining). The files (a fiction book, a news file, an object file, a paper and a C program) are described in Section 4.17.2.

In practice, the extra compression obtained by Markov methods is usually not worth the decrease in speed.[65] Fiala and Greene[Fiala89] describe some sophisticated variants on LZ77 and LZ78 and make a strong case for the practical advantages of dictionary techniques over Markov techniques. However, the field is by no means stable and it is possible that faster Markov techniques will appear (Appendix D). Meanwhile the debate continues ([Langdon83], [Langdon84], **[Hamaker88]**, [Witten88], [Fiala89]).

In summary, all current text compression techniques operate entirely by exploiting the correlations between adjacent instances. The power of a technique is measured by the number of instances used to predict

---

[63] This graph was generated based on results given in table B-1 of [Bell89]. The author of this thesis has not implemented any LZ algorithms.

[64] See figure 9–10 of [Bell89].

[65] Although the slower techniques will always find application where the cost of the channel is high relative to the cost of processing time. Examples: Modem communication, batch file compression[**Witten88**](p. 1140) and space probes.

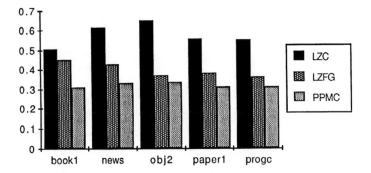

This histogram compares the compression performance of the best Ziv and Lempel algorithm (LZFG) with the best Markov algorithm (PPMC) on a number of files. The LZC algorithm, which is the Unix *compress* program, is a variant of LZ78 and is included for comparison. The vertical axis measures compression (proportion remaining) with a higher bar indicating *worse* compression than a lower bar. The horizontal axis gives the names of the test files. These are described in Section 4.17.2. This graph indicates that the best Markov algorithm yields better compression than the best dictionary algorithm.

Figure 23: Performance of dictionary and Markov techniques.

the next instance. Markov techniques apply roughly the same amount of power to each instance. Dictionary techniques apply power in a sawtooth function. Markov models yield the best compression. The best dictionary techniques yield nearly as good compression but run much faster.

1.14 Signal Compression

Data comes in two different kinds: text data and signal data. This thesis is concerned almost exclusively with text data. It is worth distinguishing between the two kinds of data because completely different techniques are used to compress them.

Text data consists of a stream of instances of symbols. Examples of text data are English text and computer programs.

Signal data consists of a stream of numbers (also confusingly called samples) taken from a real world sampling device such as a digital microphone or a video camera. Examples of signal data are digital representations of music and digital images. Each instance in a stream of signal data is a snapshot of the amplitude of an analogue source at a particular position or time.

Both text data compressors and signal data compressors compress streams consisting of instances of a set of n symbols. Each stream can equally well be thought of as a sequence of numbers in the range $[0, n - 1]$. The difference between the two kinds of data is that signal data assumes an *ordering*[66] on the symbols whereas text data does not. If text and signal messages were each fed through a one-to-one random permutation function $f : A \rightarrow A$, the text compressor would yield *identical* compression whereas the signal compressor's performance would be substantially reduced. Signal compression techniques are based on the assumption that the instances are the result of a (slowly varying) physical process.

The techniques used for modelling signal data are completely different from those used to model symbol data. However, the recent advances in symbol compression have resulted in an overlap between the fields and it is worth briefly reviewing the area. There are two interesting classes of signal data: one dimensional and two dimensional. We will refer to these classes of data as "sounds" and "images".

Two classes of technique are used to compress signal data. One method is to fit a curve to the next few instances to be transmitted and transmit only the coefficients of the curve. This has been done with Fourier transforms. Turner[**Turner75**] discusses the use of the Hadamard transformation. Shapiro[**Shapiro80**] describes a technique which dynamically selects between a number of different curves in a library of curves (e.g. linear curves, quadratic curves, exponential curves).

The second (and most popular) signal compression technique is prediction; in fact prediction was used in signal compression long before it began to star in text compression. The technique of **adaptive linear prediction**[**Witten80**] uses a linear function of the previous k instances to predict the next instance. The arithmetic difference (error) between the predicted symbol and the symbol of the actual instance, is transmitted using a smaller number of bits.

Markov models are generally ineffective when applied to signal data because the patterns in the signal data are not exact. A wave can cycle one hundred times without using the same symbol (e.g. if the symbol set size is $n = 65536$ as it is for sound on compact disks). However, Markov

---

[66] The values ordered are usually linearly distributed but are sometimes exponentially distributed (fixed point vs floating point).

models can be applied to black and white images (which have a binary symbol set $(n = 2)$) and to predicting the rough *error* of a distribution.

One of the earliest predictive techniques for compressing black and white images is described by Kobayashi and Bahl[**Kobayashi74**]. Compression takes places in two stages. Stage one does not effect a compression but simply changes the statistical properties of the block of bits. Stage two does the actual compression. In stage one, the image is scanned line by line, pixel by pixel, from left to right. Before each pixel is scanned, linear prediction is used to predict whether it will be 0 or 1. The prediction is based upon neighbouring pixels that have already been scanned. Each pixel bit is then replaced by an error bit being 0 if the prediction was correct and 1 if it was incorrect. The result is a bit matrix of the same size as the original image, but one that can reasonably be modelled as the output of a memoryless binary source. This is then encoded using binary run length coding[Bahl74] (Section 1.5.1.1). Bahl and Kobayashi discuss a number of pixel predictors and both fixed and adaptive predictors are examined.

Bahl and Kobayashi's technique can be considered as a primitive version of a more modern technique described by Langdon and Rissanen [Langdon81]. Where, in the second stage, Bahl and Kobayashi employed run length coding, Langdon and Rissanen employed a binary arithmetic code (Section 1.9.4). Where in the first stage, Bahl and Kobayashi used an adaptive linear predictor, Langdon and Rissanen used a Markov predictor. Use of a Markov predictor was possible because there were only two symbols. 1024 contexts were constructed corresponding to the 1024 different states that 10 pixels near the pixel being predicted could be in.

The other application of modern techniques to signal compression is in predicting the *error* of the prediction of an instance.

Traditionally, a signal k bits wide is compressed by predicting it and transmitting the error as a stream of instances e bits wide. If the error is greater than the maximum transmittable error, the signal becomes distorted (slew rate distortion) or an escape code is used to introduce a wider error code. A more general technique is to assume an error distribution and transmit the error according to its probability as specified by the distribution (using an arithmetic code). Todd, Langdon and Rissanen[**Todd85**] describe a technique that goes one step further by using a Markov model to predict the error yielded by a linear prediction technique. A first pass divides the error distribution into five zones called buckets, each of which contains roughly the same number of error instances. To these zones is assigned an error-symbol alphabet of size 5 that is distinct from the signal alphabet. The error symbols of the three pixels adjacent to the pixel being predicted are used to select from one of 125 contexts. The selected context predicts the error symbol of the predicted pixel. Surprisingly it was found that the actual predictor used didn't matter much if the resulting errors were being compressed by this scheme.

There are a number of ad-hoc techniques for compressing image data that do not use prediction. An image can be represented by a quad tree. Quad trees recursively divide the image into four parts, stopping when each part is a constant value. Quad trees rely on the fact that images tend to contain large tracts of the same colour. Another technique is to transmit a chain of points being the outer boundaries of a contiguously coloured area[**Morrin76**].

One of the more interesting uses for prediction in image compression is that of **progressive transmission**. Witten and Cleary[**Witten86**] describe a technique in which the image is represented by a solid, uniformly 4-furcated (quad) tree with the leaves corresponding to the pixels. The image is transmitted layer by layer from the root to the leaves, using the current layer to predict the next layer. The technique transmits the image in the same time as a single linear scan but provides coherent images of increasing resolution after each layer has been transmitted. Without using one layer to predict the next, layer by layer transmission would take about 30% longer to transmit than the leaf layer alone.

One important area of image compression is facsimile compression. Facsimile compression has become specialized because of the highly specific nature of images transmitted; typically the images are of text documents. This fact is used to advantage in the "Combined Symbol Matching Facsimile" algorithm described by Pratt, Capitant, Chen, Hamilton and Wallis[**Pratt80**]. The algorithm first scans the image looking for characters and divides the document into two bitmap overlays, one containing symbols and the other containing the residue. A character font dictionary is transmitted followed by a compressed form of the symbol portion of the page. The residue is transmitted using run-length coding. The result is an algorithm that out-performs run length coding for pages that contain a lot of text and as well as the best run length coding schemes for pages that consist mainly of graphics.

1.15 Measures of Compression

Researchers in the field of data compression seem to be uncertain of how to express the performance of their compression algorithms. If α is used to denote the length of the message and β is used to denote the length of the compressed message, a measure can be expressed in terms of α and β. Because each formula involves a division, the result of the formula is dimensionless and the units of information of α and β do not matter so long as they are the same. An informal sample of the literature revealed a high entropy in the measures used to present experimental results (**Table 4**).

Compiling this table was tedious because most papers do not explicitly define the measure that they are using. For most papers the measure used could only be determined by looking in the discussion section and locating a sentence that compares two results. The problem of identifying the measure was particularly acute when the compression presented was about 50%.

The measures listed in **Table 4** are not exhaustive. There is $\alpha/(8\beta)$ which is instances per bit and $1 - (\beta/\alpha)$ which is the proportion taken off the input file. Similar measures can be devised for nats.

Perhaps the real problem is that terms such as "compression ratio" are not descriptive of the formulas they represent[67]. At this stage, proposing standardized meanings to the various terms would be of little use. A better solution is for researchers to report their results using more precise descriptions (**Table 5**).

Nearly all of the measures listed in **Table 4** express compression in terms of the ratio of the length of the compressed file to that of a *byte stream representation* already *chosen* by humans (e.g. ASCII). These measures are relative by definition (Section 1.1.1). The ideal measure would specify the *absolute* amount of information that a technique requires to represent a given *abstract information object*. When compressing bananas, compression could be measured in information per banana where the information could be measured in digits of a given base (preferably base e). The measure "bits per instance" is the best measure proposed to date because it does not assume a representation for instances. The measure "nats per instance" is even better because, in addition, it forces the user to realize that information is a continuous quantity.

---

[67] A similar problem with nomenclature arises with Markov algorithms (Section 1.10.6).

Formula	Description
β/α	Proportion remaining
$1 - (\beta/\alpha)$	Proportion removed
$100(\beta/\alpha)$	Percentage remaining
$100(1 - \beta/\alpha)$	Percentage removed
$8(\beta/\alpha)$	Bits per instance
$\alpha/(8\beta)$	Instances per bit
$8(\ln 2)(\beta/\alpha)$	Nats per instance
$\alpha/(8(\ln 2)\beta)$	Instances per nat
α/β	Compression gain

One method of improving the clarity of compression results is to use more descriptive terms for compression rather than generic terms such as "compression". This table lists proposed terminology for various compression measures.

Table 5: Proposed nomenclature for performance measures.

and cryptography. Error correction provides a noiseless channel without which data compression would be hazardous. Data compression enhances the security of cyphers. Cyphers form a dual with error correcting codes. This section describes this cycle.

Conventional data compression techniques provide a simple one-to-one mapping between messages and code words. If a portion of the coded message is corrupted, the text can usually be recovered manually. In contrast, modern data compression techniques introduce such a complex of dependencies that the influence of a single instance can propagate throughout the rest of the coded message. This makes errors much harder to correct. Data compression relies on the presence of a noiseless communication line which can be approximated only with the use of error correction. Data compression and error correction systems are symmetric with respect to redundancy; data compression removes redundancy whereas error correction introduces it.

The link between data compression and cryptography is nearly as strong. At a purely practical level, the introduction of data compression into a communications system provides an extra level of complexity which the cryptanalyst must work through[**Rubin79**]. Anyone who has tried to program a finite-context predictive model driving an arithmetic code will realize how sensitive to detail the whole system is. It would be difficult to find the exact system being used by an enemy even if data compression were the only "cypher" being used. Cryptographers place no weight on this argument, however, as they assume that the cryptanalyst has access to the cypher algorithm, including protocols and data compression layers. Cryptanalysts define three levels of attack. If only the cyphertext is known, the attack is called a **cyphertext only attack**. If a piece of text and its cyphertext is known, the attack is a **known plaintext attack**. If the cryptanalyst has the capability to inject an unlimited number of messages into the cypher and observe the resulting cyphertext, it is a **chosen plaintext attack**. The last threat is the one most commonly addressed by cryptographers.

Early theoretical work on "Secrecy Systems" was performed by Shannon[**Shannon49**] who used his newly founded field of information theory to provide a solid foundation for cryptography. Shannon modelled a cypher as a mapping from a set of messages to a set of codewords (**Figure 24**). The mapping is many to many. The cypher key (which the cryptanalyst does not know) resolves the ambiguity. Shannon constructed a measure of the security of a cypher based on the average number of messages which map into an arbitrary cyphertext. This enabled him to prove that there is such a thing as an unbreakable cypher, the existence of which was an open question at the time.

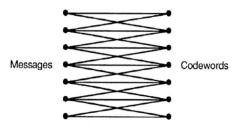

Shannon modelled a cypher as an ambiguous many-to-many mapping between messages and codewords that could only be resolved using the cypher key.

Figure 24: Shannon's model of cryptography.
(Based on figure 4 and figure 5 of [Shannon49])

With the increasing need for fast, reliable, secure computer networks, cryptography is again in the spotlight. In an excellent paper, Diffie and Hellman[**Diffie76**] reviewed the current and future requirements for cryptographic systems, and then introduced public key cryptography. The paper "caused a basic revolution in the way people think about cryptographic systems"[**Tanenbaum81**](section 9.1.4) because no-one up to that point had considered it possible to form a secure channel without a prior secret key exchange. Diffie and Hellman also discussed the relationships between various problems in cryptography and ended with a discussion of complexity theory[**Garey79**].

In a later paper, Hellman[**Hellman77**] discussed Shannon's approach to cyphers. In particular, Hellman emphasised a point made by Shannon about the importance of data compression in cryptographic systems. Data compression techniques map a large space of redundant source messages[68] into a smaller space of less redundant messages. If the message is encrypted *after* compression, the ambiguity of the resulting cyphertext is increased because fewer of the possible decyphered messages are meaningless (**Figure 25** and **Figure 26**).

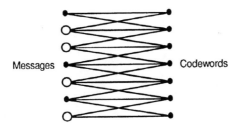

Without data compression, the set of sensible messages is sparse. This means that for any given codeword, there will be less possible *sensible* (non-hollow) messages that it could correspond to. This decreases security. Here, the fourth code word could only correspond to the fourth message because messages three and five are non-sensible.

Figure 25: Encryption without data compression.
(Based on figure 4 of [Hellman77])

At the end of his paper, Hellman completes the circle by connecting cryptography and error correction in a dual relationship. In the absence of "source coding" (a common term for data compression among theorists), the best error correction code is one that provides a random mapping from codewords to messages (so as to include as many meaningless messages as possible) whereas the best cypher is one that provides a non-random mapping (so as to include as many meaningful messages as possible). This view contrasts with Shannon's view which states that a random mapping is optimal in both cases.

---

[68] In this case the term "redundant" is used to mean that the space of messages contains many messages that will *never* be sent.

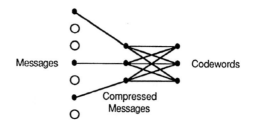

With data compression, the non-sensible messages are eliminated before encryption takes place. As a result, code word ambiguity increases.

Figure 26: Encryption with data compression.
(Based on figure 3 of [Hellman77])

To summarize, error correction, data compression and cryptography form a tight triangle. In particular, error correction is necessary for data compression to be feasible over noisy channels, and data compression is fundamental to providing cryptographic security.

1.17 Summary

Data compression has been in use for hundreds of years. The introduction of information theory, computers and communication networks has greatly accelerated the field and added urgency to its application. Whereas information is a continuous quantity, computers can only store a finite set of information sizes, the smallest size being one bit. This makes the storage of small units of information inefficient. Huffman coding, the all-time most popular technique, suffers from this malady. Early attempts to overcome this inefficiency involved mapping blocks of source symbols to blocks of channel symbols with the aim being to minimize the difference in information content.

Following the introduction of Huffman coding, the field stagnated for thirty years, during which a number of ad hoc techniques were developed. Principal among these were run length coding and the dictionary techniques.

In the late seventies, the field warmed up again. Dictionary techniques became adaptive. Arithmetic coding emerged as the solution to the coding inefficiency problem. Two theorems arose in the early eighties consolidating these ideas. The first asserted the superiority of single-instance parsing over multiple-instance parsing. The second asserted the superiority of one-pass adaptive compression over two pass semi-adaptive compression. These two ideas along with arithmetic coding formed the modern paradigm of data compression from which finite-context Markov algorithms arose. Finite-context Markov algorithms predict each instance based on the previous few instances of the message and on the past behaviour of the source.

Adaptive dictionary (Ziv and Lempel) techniques are the most practical, operating at high speed and giving good compression. Markov models operate more slowly but yield better compression.

Chapter 2

The DHPC Algorithm

2.1 Introduction

Chapter 1 contained a review of the development and current status of data compression. The remainder of the thesis presents the results of the author's work in this field.

In this chapter we present a finite-context algorithm called **Dynamic History Predictive Compression** (DHPC) that was devised by the author in late 1986 and published in 1988[Williams88]. DHPC is similar to PPM in its basic approach but differs in all other aspects. The algorithm will be described in detail because it is

- an original contribution.

- a concrete example of a Markov algorithm.

- the basis for more sophisticated algorithms.

DHPC is a variable-order finite-context Markov algorithm which conforms with the modern paradigm. It uses a backwards tree to store its contexts and grows the tree incrementally as compression proceeds. Each node in the tree contains a sample which is updated whenever the node matches the history. Predictions are based solely on the deepest matching node that contains enough instances to be reliable.

2.2 Tree Growth

DHPC's tree growth is achieved by adding leaf nodes, one at a time, at a maximum rate of one node per instance. Growth is ultimately bounded by a predetermined maximum $z \in \mathbf{Z}[1, \infty)$ on the number of nodes and a predetermined maximum $m \in \mathbf{Z}[0, \infty)$ on the depth of a node. Tree growth continues forever only if $m = \infty$ and $z = \infty$. Once a node is attached to the tree, it is never moved; once the tree contains z nodes, its structure does not change. It is therefore important to build the tree in a manner that ensures that its final structure approximates that of the source.

DHPC controls the *rate* of growth by preventing growth from nodes whose samples are not *extensible*. A node is **extensible** if its sample contains at least α instances where $\alpha \in \mathbf{Z}[2, \infty)$ is a constant parameter of the algorithm called the **extensibility threshold**.

The rules for tree growth are summarized as follows. After each instance has been transmitted and used to update the matching branch, a new matching leaf node is attached to the deepest matching node (u — the potential parent node) in the tree unless one or more of the following conditions hold.

- The tree contains z nodes.

- u is of depth m.

- u is not extensible.

This policy ensures that the rate of growth of each part of the tree is proportional to its frequency of access. This increases the likelihood that the eventual structure of the tree will approximate that of the source. The algorithm attempts to construct a tree that maximizes the average access depth so as to maximize the average order of finite-context Markov model used to make predictions.

2.3 Estimation

The DHPC algorithm produces a prediction for each incoming instance. Each prediction is based solely on the sample of the deepest matching credible node. A node is **credible** if its sample contains at least β instances where $\beta \in \mathbf{Z}[1, \infty)$ is a constant parameter of the algorithm called the **credibility threshold**.

DHPC converts the sample of the deepest matching credible node into a prediction using linear estimation (Section 1.10.3.1) with $\lambda = 1$. That is

$$\xi(x)(a) = \frac{x(a) + \frac{1}{n}}{y + 1}$$

yielding the estimated probability of a symbol a recorded $x(a)$ times in a sample of y instances ($y = \sum_{a \in A} x(a)$). The low value of λ indicates that the algorithm places a lot of trust in the samples chosen for prediction, those samples having to contain at least β instances.

2.4 The Algorithm in Detail

The DHPC algorithm manipulates two data structures: a backwards tree and a sliding-window history buffer. The backwards tree models the source. The sliding window history buffer stores the m most recent instances of the history.

Four constant parameters control the behaviour of the algorithm.

The **depth limit** $m \in \mathbf{Z}[0, \infty)$ defines the maximum depth of the tree and hence also the length of the history buffer. The depth limit places a constant upper bound on the time needed to process each instance and a weak ($O(n^m)$) upper bound on memory consumption.

The **node limit** $z \in \mathbf{Z}[1, \infty)$ defines the maximum number of nodes in the tree. The tree structure is frozen when it reaches z nodes. The node limit places an upper bound on memory consumption.

The **extensibility threshold** $\alpha \in \mathbf{Z}[2, \infty)$ defines the minimum number of instances that a node's sample must contain before the node can sprout leaf nodes.

The **credibility threshold** $\beta \in \mathbf{Z}[1, \infty)$ defines the minimum number of instances that a sample must contain for it to be used to make a prediction.

A review of previous definitions may resolve some ambiguities. The depth of a node is defined to be the number of arcs between the node and the root node. The root node is defined to be of depth zero. A tree containing only the root node contains one node. y is an abbreviation for $\sum_{a \in A} x(a)$. A node is extensible if its sample contains at least α instances. A node is credible if its sample contains at least β instances. The terminology of extensibility and credibility applies to samples as well as nodes.

2.4.1 Main Program

The main program of DHPC is listed in **Figure 27**. The algorithm is best cast as a process that reads a stream of instances and writes a stream of predictions. This enables it to be slotted into the model unit of **Figure 9**.

In this formulation, the history buffer is represented by an array *hist* of m instances with *hist*[1] containing the most recently received instance. The tree is stored in some sort of dynamic data structure and is accessed through a "pointer" called *root* which points to the root node. Each node u contains a sample x of n frequency counts $u.x(a)$, $\forall a \in A$.

```
process DHPC(in instancestream; out predictionstream) is
    α : constant integer ← <Extensibility threshold parameter>;
    β : constant integer ← <Credibility threshold parameter>;
    m : constant integer ← <Depth limit parameter>;
    z : constant integer ← <Node limit parameter>;
    type history is array[1 ... m] of symbol;
    <Procedures predict and update>
begin DHPC
    root : node;
    hist : history;
    new(root); root.x[a₁ ... aₙ] ← 0;
    hist[1 ... m] ← ' ';
    loop
        instance : symbol;
        write(predictionstream,predict(root,hist));
        read(instancestream,instance);
        update(root,hist,instance);
    end loop;
end DHPC;
```

The DHPC main program implements the model unit of **Figure 9**, reading a stream of instances and generating a stream of predictions. The algorithm generates a prediction for each instance before it reads the instance. The *predict* function and *update* procedure do all the work.

Figure 27: DHPC main program.

The algorithm starts by initializing the tree (which is manipulated through its root node *root*) to a single (root) node and filling the history buffer (*hist*) with spaces.[69] Each instance is processed in two phases: a **prediction phase** during which a prediction for the next instance is generated, and an **update phase** during which the next instance is used to update the history buffer and the tree. *predict* is a function with no side effects. *update* is a procedure that modifies its parameters but has no other side effects. The next two sections describe the subprograms *predict* and *update* that implement these phases.

2.4.2 Prediction Phase

The prediction phase produces a prediction of the next instance. This phase is controlled by function *predict* (**Figure 28**). The function does not modify its arguments and has no side effects.

function *predict*(**in** *root* : *node*; **in** *hist* : *history*) : *prediction* **is**
begin *predict*
   if *root.y*$< \beta$ **then**
      return *p*: $p \in P \land \forall a \in A,\ p(a) = \frac{1}{n}$;
   end if;
   current : *node* $\leftarrow$ *root*;
   loop
      exit if *depth*(*current*)=*m*;
      next : *node* $\leftarrow$ *current.child*[*hist*[*depth*(*current*)+1]];
      exit if not *exists*(*next*);
      exit if *next.y*$< \beta$;
      current$\leftarrow$*next*;
   end loop;
   return *p*: $p \in P \land \forall a \in A,\ p(a) = \xi(current.x)(a)$;
end *predict*;

The DHPC algorithm bases each prediction entirely on the sample of the deepest matching credible node. If no such node exists, the uniform prediction is returned.

Figure 28: DHPC prediction function.

---

[69] Because $\alpha > 1$, history length always exceeds tree depth, making this initialization strictly unnecessary.

Function *predict* starts by examining the root node and returning with the uniform prediction if the root is not credible. It is an invariant that if the root is not credible, all nodes are not credible.

If the root is credible, *predict* begins at the root and moves down the matching branch stopping on the deepest credible node. The node's sample is then fed through the estimation function ξ (Section 2.3) producing the prediction.

2.4.3 Update Phase

Between the prediction and update phases, the main program (**Figure 27**) reads in the instance corresponding to the prediction made during the prediction phase. The update phase uses the new instance (*instance*) to update the history and the tree. This phase is controlled by procedure *update* listed in **Figure 29**. The procedure modifies its first two parameters but has no other side effects.

Procedure *update* moves down the matching branch adding the new instance to the sample of each node visited. It stops on the deepest matching node u. If, at that point, u is not of maximum depth, and u is extensible, and the tree contains less than z nodes, a matching child node (a leaf) is created and initialized with an empty sample. The new instance is added to the new node's sample as if the new node had been part of the matching branch all along. Once this is done, the instance is shifted into the history, the oldest instance in the history being discarded.

Thus a sample receives its first instance upon the creation of its node and collects one instance each time its node matches the history. Because $\alpha > 1$, at most one node is added to the tree for each instance read.

The statement **exit if** *isleaf*(*current*) exists only to prevent an illegal access of $hist[m+1]$ on the following line and can be eliminated by declaring *hist* with an index range of $\mathbf{Z}[1, m+1]$.

2.5   Example Execution of DHPC

We now present an example execution of DHPC. The example uses a three symbol alphabet ($n = 3$) consisting of the symbols a, b and c. The maximum depth of the tree is set to six arcs ($m = 6$) and the maximum number of nodes to one hundred ($z = 100$). The extensibility threshold is set to three instances ($\alpha = 3$) and the credibility threshold to two instances ($\beta = 2$). These parameters are summarized in **Table 6**. The message starts with abcbacbabc.

procedure *update*
    (**in** *root* : *node*; **in out** *hist* : *history*; **in** *instance* : *symbol*);
begin *update*
    current : *node* ← *root*;
    loop
        inc *current.x*[*instance*];
        exit if *isleaf*(*current*);
        exit if not *exists*(*current.child*[*hist*[*depth*(*current*)+1]]);
        current ← *current.child*[*hist*[*depth*(*current*)+1]];
    end loop;
    if (*depth*(*current*)< *m*) **and** (*treesize*(*root*)< *z*) **and**
    (*current.y* ≥ *α*) **then**
        newnode : *node*;
        new(*newnode*); *newnode.x*[$a_1 \ldots a_n$] ← 0;
        current.child[*hist*[*depth*(*current*)+1]] ← *newnode*;
        inc *newnode.x*[*instance*];
    end if;
    for *i* **in reverse** 2 . . . *m* **loop** *hist*[*i*] ← *hist*[*i* − 1] **end loop**;
    hist[1] ← *instance*;
end *update*;

After each instance has arrived, the DHPC algorithm adds it to the sample of each node on the matching branch (first **loop**), working from root to tip. When it reaches the end of the branch, it sometimes adds an extra node on the end (**if**). The **for loop** slides the history buffer.

Figure 29: DHPC update procedure.

Parameter	Value
n	3 : (a, b, c)
m	6
α	3
β	2
z	100

Table 6: Parameters used in the DHPC example.

Execution of the algorithm will be illustrated using a sequence of snapshots that summarize the state of processing between the arrival of each instance. The snapshots are taken between the *write* and the *read* statements of the main loop of the main procedure listed in **Figure 27**.

A snapshot labelled "Instances read : k" was taken at the time when the *read* statement had been executed exactly k times.

Each snapshot gives a picture of the tree, the number of instances already read, the position in the input string, the history buffer, the sample used to make the next prediction, and the next prediction itself. The tree is drawn with leaves to the left so as to align it with the history through which instances flow from right to left in accordance with Western reading conventions. Each node in the tree is labelled by its sample in the form of a triple followed by the letter 'C' if the node is credible and by the letter 'E' if the node is extensible. Instances to the left of the vertical bar in the input string have already been read. The array *hist* is displayed with *hist*[1] at the rightmost end in order to illustrate the right to left flow of instances. Full stops take the place of spaces. The sample used to make the next prediction is displayed as a triple containing the counts for each symbol, in the order (a, b, c). This sample when fed into ξ yields the next prediction which is expressed as a vector of rational probabilities obtained from the form $\xi(x)(a) = \frac{nx(a)+1}{ny+n}$. This form is derived from the form given in Section 2.3 by multiplying the numerator and denominator by n.

At time 0, no instances have been processed. This snapshot illustrates the state of the data structures after the initialization but before the first instance has been read. The tree consists of the root node, whose sample is empty. The history is filled with spaces. Despite the lack of information, a prediction must be made. In the absence of a credible node, the uniform sample of $(1,1,1)$ is used, resulting in a uniform prediction of $(4/12,4/12,4/12)$.

$$\bullet^{(0,0,0)}$$

```
Instances read      : 0
Input string/posn   : "|abcbacbabc"
History buffer      : "......"
Prediction sample   : (1,1,1)
Prediction          : (4/12,4/12,4/12)
```

Instance 1, which is a, arrives and is added to the samples of the matching branch, which consists of the root node. The root node is not extensible so no growth occurs. The history buffer is shifted. As there is no credible node, the prediction is based on the uniform sample.

●(1,0,0)

```
Instances read    : 1
Input string/posn : "a|bcbacbabc"
History buffer    : ".....a"
Prediction sample : (1,1,1)
Prediction        : (4/12,4/12,4/12)
```

Instance 2, which is b, arrives and is added to the samples of the matching branch, which consists of the root node. The root node, which now contains two instances, is credible but not extensible, so no growth occurs. The history buffer is shifted. As the root is credible, it so it is used to make the prediction. The root node's sample of (1,1,0) results in the prediction (4/9,4/9,1/9). The estimation function ξ ensures that the third symbol (c) is allocated a small probability despite its zero frequency.

●(1,1,0)C

```
Instances read    : 2
Input string/posn : "ab|cbacbabc"
History buffer    : "....ab"
Prediction sample : (1,1,0)
Prediction        : (4/9,4/9,1/9)
```

Instance 3, which is c, arrives and is added to the samples of the matching branch, which consists of the root node. The arrival of this third instance makes the root extensible (because $3 = root.y \geq \alpha = 3$), allowing it to grow a new matching child node. The new node is created at depth $d = 1$ with an arc labelled b so as to match the history $hist[d]$. The new node's sample is updated as if it had been part of the tree all along. This results in a sample of (0,0,1). The history buffer is shifted. The root is the deepest credible matching node and so its sample of (1,1,1) is used to make the prediction.

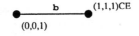

(0,0,1)

```
Instances read    : 3
Input string/posn : "abc|bacbabc"
History buffer    : "...abc"
Prediction sample : (1,1,1)
Prediction        : (4/12,4/12,4/12)
```

Instance 4, which is b, arrives and is added to the samples of the matching branch which consists of the root node. The deepest node in the matching branch is the root node, which is extensible, and so a new child node is created whose arc is labelled c. The new node is updated with the new instance (b). The history buffer is shifted. At this point the deepest matching node is the node on the b arc, but as it is not credible, the root node's sample is again used to make the prediction.

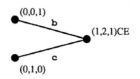

```
Instances read     : 4
Input string/posn  : "abcb|acbabc"
History buffer     : "..abcb"
Prediction sample  : (1,2,1)
Prediction         : (4/15,7/15,4/15)
```

Instance 5, which is a, arrives and is added to the samples of the matching branch which consists of the root node and node b. This update makes the node b credible. However node b is not yet extensible and so no new node is added. The history buffer is shifted. As there is no node a, the root is the deepest credible matching node and so it is used to make the prediction.

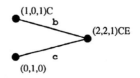

```
Instances read     : 5
Input string/posn  : "abcba|cbabc"
History buffer     : ".abcba"
Prediction sample  : (2,2,1)
Prediction         : (7/18,7/18,4/18)
```

Instance 6, which is c, arrives and is added to the samples of the matching branch which consists only of the root node. As the root node is extensible, a new child node a is created. The history buffer is shifted. As node c is not yet credible, the root's sample is used to make the prediction.

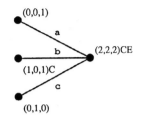

```
Instances read      :  6
Input string/posn   :  "abcbac|babc"
History buffer      :  "abcbac"
Prediction sample   :  (2,2,2)
Prediction          :  (7/21,7/21,7/21)
```

Instance 7, which is b, arrives and is added to the samples of the matching branch which consists of the root node and node c. Node c is not yet extensible and so no growth takes place. The history buffer is shifted. The prediction is based on the sample of node b which became credible when instance 5 arrived.

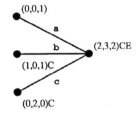

```
Instances read      :  7
Input string/posn   :  "abcbacb|abc"
History buffer      :  "bcbacb"
Prediction sample   :  (1,0,1)
Prediction          :  (4/9,1/9,4/9)
```

Instance 8, which is a, arrives and is added to the samples of the matching branch which consists of the root node and node b. This makes the node b extensible and so a new matching child node cb is created. The new instance is added to the new node's sample as if it had been in the tree all along. The history buffer is shifted. The deepest credible matching node is the root node because node a is not yet credible.

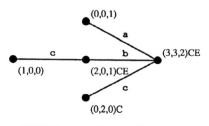

```
Instances read      : 8
Input string/posn   : "abcbacba|bc"
History buffer      : "cbacba"
Prediction sample   : (3,3,2)
Prediction          : (10/27,10/27,7/27)
```

Figure 30 depicts the state of the algorithm's data structures after a further 34 instances have been processed. As neither the depth nor the node limit has been reached, all extensible nodes are non-leaf nodes. The prediction of $(4/12, 7/12, 1/12)$ is derived from the sample $(1,2,0)$ of node ac. Node bac matches but is not yet credible.

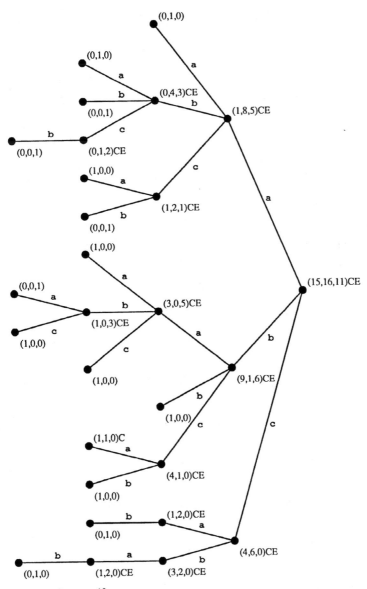

Instances read   : 42
Input string/posn   : "abcbacbabcabcababcacaababcbacbbacbababcbac|"
History buffer   : "abcbac"
Prediction sample   : (1,2,0)
Prediction   : (4/12,7/12,1/12)

Figure 30: DHPC tree structure after 42 instances.

2.6   A Comparison of DHPC with Other Algorithms

DHPC can be considered to be a generalization of the DAFC algorithm[Langdon83] described in Section 1.10.6.2. The DAFC algorithm constructs a fractional order model by starting with a single general context (whose string is the empty string) and then constructing up to 31 first order contexts. A first order context a is created when γ (a constant parameter of DAFC) instances of the symbol a have been recorded in the general context. For the purposes of comparison, DAFC's run-length feature and optimized estimation technique are ignored.

The DHPC algorithm with settings $m = 1$, $z = 32$, $\alpha = \gamma$ and $\beta = 1$ is almost identical to DAFC. The major difference is that DHPC creates a child node if the *parent* node contains more than a certain number of instances, whereas DAFC creates a child node if the *child* node contains more than a certain number of instances. DAFC accomplishes this seemingly impossible feat by recording how often each child *would* have been accessed had it existed. It does this without using extra memory by examining the frequencies in the zero order sample which happen to be identical to the first order context frequencies.[70]

Despite their differences, DHPC and DAFC's similarities betray their identical design goals of increasing order while conserving memory. Both use instance count thresholds to determine the rate of growth in the hope of maximizing the average access depth. Both base their predictions on the sample of a single node. DHPC is really a recursive DAFC.

The relationship between DHPC and PPM is an interesting one. PPM sets a depth limit m, but otherwise does not address the memory problem. PPM creates a context for every string of length m or less in the history. This is the same as setting $\alpha = 1$ in DHPC.[71]

DHPC and PPM differ significantly in their approach to estimation. DHPC chooses the deepest matching node that has at least β samples, and uses linear estimation. PPM uses the deepest matching node (u), and uses non-linear estimation (Section 1.10.3.2), but uses the parent of u's sample (less instances of all symbols appearing in u) to divide the probability represented by the $\lambda/(y + \lambda)$ term among the zero-frequency

---

[70] It may be possible to extend this technique to many levels but it is not immediately clear how this could be done.

[71] DHPC does not allow this but could easily be modified to allow $\alpha = 1$ by moving the **if** statement of **Figure 29** into the loop above it.

symbols (**Figure 15**). This division continues recursively, terminating at the order -1 sample. Both DHPC and PPM set $\lambda = 1$.

To summarize, PPM uses the deepest node no matter how few samples it has, but is careful in allocating the zero frequency probabilities. DHPC is careful to select a node with a credible number of samples but allocates equal probability to the zero frequency symbols.

Experimental results indicate that the PPM estimation technique performs about 7% absolute better than DHPC (Experiment 2 in Section 4.17.5). However, DHPC does not blend the samples of different levels of the matching branch. This allows it to use more efficient representations for predictions (Section 1.11).

On the implementation side, PPM is implemented using a forwards tree whereas DHPC is implemented using a backwards tree.

2.7 An Analysis of Tree Growth

One of the innovations of DHPC is the use of an extensibility threshold to control the rate of tree growth. This section presents upper and lower bounds for the rate of tree growth of DHPC.

We assume that unbounded resources are available ($m = \infty$, $z = \infty$) and consider the amortized constraints on attributes of the growing tree. The concept of amortized computational complexity has recently been introduced by Tarjan and Sleator[Tarjan85][**Sleator86**] as a way of obtaining tighter complexity bounds for sequences of operations. Traditional complexity measures consider the worst case for each operation in a sequence; amortized complexity measures consider the worst *sequence*.

Theorem 1: For $\alpha > 1$, $m = \infty$, $z = \infty$, DHPC grows its tree at a minimum rate of one node per $\alpha - 1$ instances and a maximum rate of one node per $((\alpha - 2)/n) + 1$ instances.

Proof: We consider a particular tree at a particular time and define two integer attributes a and e. Attribute a is the maximum number of consecutive instances that the tree could absorb without causing tree growth. It is defined as

$$a = \sum_{(i \in Nodes) \wedge (i.y < \alpha - 1)} (\alpha - 1) - i.y$$

Attribute e is the maximum number of consecutive instances that would cause new tree growth. It is defined as

$$e = \sum_{(i \in Nodes) \wedge (i.y \geq (\alpha - 1))} n - children(i)$$

These definitions assume that each instance has the liberty of "choosing" to arrive at any node. In fact, the set of nodes that can be chosen is severely constrained by the previous few instances. This assumption does not weaken the proof but it does mean that tighter bounds might exist.

For the purposes of the proof, the arrival of an instance is viewed as causing one of two operations to be performed depending on whether the arrival of the instance causes a new node to be added to the tree. If a new node is added, the *Add* operation is performed. If no new node is added, a *Hit* operation is performed. The effect of each of these operations on attributes a and e is defined as follows.

$$Add: e > 0 \longrightarrow [a \leftarrow a + (\alpha - 2); \quad e \leftarrow e - 1]$$

$$Hit: a > 0 \longrightarrow \left[a \leftarrow a - 1; \quad e \leftarrow e + \frac{n}{\alpha - 2}\right]$$

The guards on these operations protect them from being performed illegally. For example, by definition, the *Add* operation cannot be performed if $e = 0$.

When a node is added, the parent of the new node has one less slot for expansion and so e is decremented. The newly created node has no child nodes. This means that it can absorb $\alpha - 2$ instances without creating a new node. Thus a is increased by $\alpha - 2$.

The *Hit* operation has a counteracting effect. The absorption of an instance decreases the capacity of the tree to absorb and so a is decremented. On the other hand, the node that was hit moves just a little closer to becoming extensible, at which time e will increase by n. This discontinuity is hard to represent within the amortization framework used here. However, the overall *rates* at which operations are performed can be represented correctly by adding $n/(\alpha - 2)$ to e each time the hit operation is performed. The justification for this is that once created, each node will absorb $\alpha - 2$ instances after which n child slots become available.

Now consider a long sequence of A *Add* and H *Hit* operations. At the start of the sequence, the tree consists of just a root node with $a = \alpha - 2$

and $e = 0$. At the end of the sequence $a, e \geq 0$. Examination of the effect of the operations yields the following constraints which give the upper and lower bounds on the ratios of A to H.

$$(\alpha - 2)A - H \geq 0 \Rightarrow \text{Min } A = \frac{H}{\alpha - 2}$$

$$\left(\frac{n}{\alpha - 2}\right) H - A \geq 0 \Rightarrow \text{Max } A = \frac{nH}{\alpha - 2}$$

Because A corresponds to the number of nodes in the tree and $A + H$ corresponds to the number of instances processed, the average growth rate is

$$Rate = \frac{A}{A + H}$$

Substituting the minima and maxima of A yields the following bounds.

$$\text{Min Rate} = \frac{\frac{H}{\alpha - 2}}{\frac{H}{\alpha - 2} + H} \qquad \text{Max Rate} = \frac{\frac{nH}{\alpha - 2}}{\frac{nH}{\alpha - 2} + H}$$

Multiplying top and bottom by the inverted top yields

$$MinRate = \frac{1}{1 + (\alpha - 2)} \qquad MaxRate = \frac{1}{1 + \frac{\alpha - 2}{n}}$$

Which completes the proof.

Figure 31 shows the minimum and maximum growth rate (in nodes per instance) plotted against α for $n = 256$. Both curves are $1/\alpha$ but the maximum decreases significantly (to 0.5) only when α approaches n. In contrast, the minimum drops rapidly. Setting $\alpha = 2$ guarantees a rate of exactly one new node per instance. Setting $\alpha = 3$ guarantees a rate of between about 0.5 and 1 new nodes per instance. Tight bounds are possible for very high values of α, but such values are inappropriate for practical purposes.

In practice, the position of the node growth between the minima and maxima curves will be determined by the entropy of the source. For ordinary sources (e.g. text files) the entropy is relatively low at depths of two or three and the growth rate is likely to be closer to the minimum than the maximum. In practice, the depth limit will prevent growth from continuing at the "hot spots" of a tree. As a result, the effective rate is likely to be quite low.

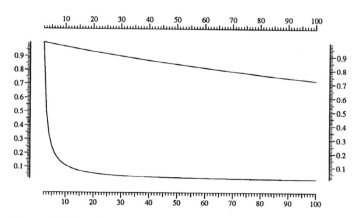

This graph plots theoretically derived lower and upper bounds for the
growth rate (nodes per instance) of the DHPC algorithm against the
extensibility threshold α.

Figure 31: DHPC minimum and maximum growth rates against α.

2.8 Summary

The DHPC algorithm employs a variable-order, finite-context Markov
model with four parameters: a depth limit m, a node limit z, an
extensibility threshold α and a credibility threshold β. The algorithm
starts with a root node and grows a backwards tree by attaching a new
leaf to the tip of the matching branch whenever a new instance arrives.
A node is not attached if the deepest matching node is of the maximum
depth m or if it contains less than α instances. The α threshold ensures
that the local rate of tree growth is proportional to the local rate of access,
resulting in a tree that maximizes the average access depth. Arriving
instances are added to the samples of all matching nodes. Each instance
is predicted using the sample of the deepest matching node that contains
at least β instances. After z nodes have been allocated, the tree structure
becomes static. At this point the tree structure should mirror that of the
source. DHPC builds its tree more slowly than the PPM algorithm and
bases its predictions on a single node rather than a blend of the nodes
on the matching branch. DHPC is faster than PPM but yields poorer
compression.

DHPC's parameterization, simplicity of growth, simplicity of estimation and simplicity of implementation make it an excellent platform from
which to explore the adaptive mechanisms of the class of variable-order,
finite-context Markov algorithms.

Chapter 3

A Classification
of Adaptivity

3.1 Introduction

So far, we have reviewed the field of data compression and presented a new Markov algorithm called DHPC. In this chapter we investigate adaptivity in data compression algorithms and describe how various kinds of adaptivity can be incorporated into DHPC. This yields new insights into what is desirable in an adaptive algorithm and into mechanisms for implementing adaptive algorithms in general.

3.2 Previous Definitions of Adaptivity

The term "adaptive" is often used, but poorly defined. The root word "adapt" is itself messy, having over 20 different forms (Appendix B). In the field of data compression, the word "adaptive" has been used loosely to describe any algorithm that varies its compression technique in response to the data.[72] Recently, the definition of the term has been tightened. One of the contributions of this thesis is to tighten the definition further.

Recent data compression literature (e.g. [Rissanen81], [Cleary84]) has defined the word "adaptive" as a term descriptive of one-pass algorithms that change the way that they compress in response to the history. This definition aligns with Assumption 5 (Section 1.6) and includes algorithms that use alphabet extension as well as those that do not.

---

[72] The word "dynamic" has been used in a similar manner.

The current definition of the word "adaptive" seems to have arisen from Rissanen and Langdon's work on the modern paradigm [Rissanen81].[73] Although Rissanen and Langdon were exponents of adaptive coding, in [Rissanen81] they were primarily concerned with alphabet extension vs single-instance prediction rather than semi-static vs adaptive techniques. It was Cleary and Witten[Cleary84] who proved the superiority of adaptive techniques over semi-adaptive techniques. However, Langdon and Rissanen considered both adaptive and non-adaptive ("static" in Cleary/Witten terminology) techniques and took the time to give the best definition of adaptivity yet.

Rissanen and Langdon defined adaptivity in terms of their formalization of models. A model is organized as a collection of **contexts** defined by a **structure function**. The structure function maps the infinite set of possible history strings onto a finite set of contexts. For example, a set of n contexts (for a first order model) might correspond to the set of conditions $h_{|h|} = a, \forall a \in A$. Associated with each context is a sample that records the instances generated by the source under the conditions defined by that context. Rissanen and Langdon defined an adaptive technique as one that (within a single context) employs a counting/estimating technique that does not decrease the estimated probability of a symbol if, within any given substring of the message, instances of the symbol occur with a higher than estimated frequency. This is a simple constraint on the way in which a model must react to the message if it is to be called adaptive.

Rissanen and Langdon's definition of adaptive is (briefly) described and accepted in two other significant papers[Cleary84][Cormack87]. Chapter 3 of a book by Bell, Cleary and Witten[Bell89] is devoted to "Adaptive Models". In it, Bell, Cleary and Witten distinguish between static, semi-adaptive and adaptive models and present the proof in [Cleary84] that shows that adaptive models are superior to semi-adaptive models.

Recently the term "locally adaptive" has been used to refer to techniques that base their prediction only on the recent behaviour of the source. An example of this usage is in the paper title "A Locally Adaptive Data Compression Technique"[Bentley86].

In this chapter the issue of adaptivity is addressed by:

- classifying forms of adaptivity.

---

[73] A brief review of Rissanen and Langdon's definition of adaptivity appears in section D of [Langdon81].

- classifying sources and the ways in which they change.

- investigating the performance of different kinds of adaptivity on different kinds of sources.

- discussing how various forms of adaptivity can be implemented.

3.3 A Classification of Adaptivity

In this section, Rissanen and Langdon's definition of adaptivity is extended to define four classes of adaptivity. The four classes cover all the data compression algorithms described to date (**Figure 32**).

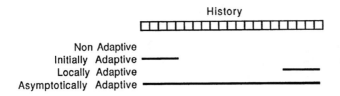

Compression algorithms can be roughly sorted into four categories of adaptivity depending on which part of the history is used to make predictions.

Figure 32: Use of the history by four kinds of adaptivity.

Non Adaptive: Does not alter its model in response to the history.

Initially Adaptive: Builds its model from a finite number of instances at the start of the history.

Locally Adaptive: Builds its model from a finite number of instances at the end of the history.

Asymptotically Adaptive: Uses all of the history to build its model.[74]

This refinement of adaptivity focuses not on the *estimation technique* but on the *location* from which a model obtains its information.

The definitions above provide a good rule of thumb but are unsatisfactorily vague. For example the definitions fail to classify a model that

---

[74] As the history length increases, the performance of an asymptotically adaptive, finite-context, order-k Markov model should converge asymptotically on the entropy of a finite-context, ergodic, order-k Markov source.

uses all but the first instance of the history, when obviously such a model should be classified as asymptotically adaptive. A more precisely stated, fuzzier definition is required.

The issues of adaptivity are greatly simplified by treating only zero-order Markov models and sources. This can be done without loss of generality by treating sources and models as collections of contexts organized by a structure function. Once the zero-order case is taken care of, higher order sources and models can be constructed from the zero order case using structure functions.

Consider a zero-order source $d(S_0)$ that produces a stream of predictions which in turn are used to produce a stream of instances. After l instances have been produced, the history is $h \in S_l$. A zero order model can be described by a tuple (w, ξ) where $\xi(x)$ is an estimation function (Section 1.10.3) and $w(l, i)$ is a **weight function** that specifies the *emphasis* that a model places on the i'th instance of a history of length l (with the first (leftmost) instance being numbered 1). The function w is normalized for a given l. That is

$$\forall l \in \mathbf{Z}[1, \infty), \ \sum_{i=1}^{l} w(l, i) = 1$$

A zero order model estimates the probability of the next instance being a as[75]

$$p(a) = \xi(x)(a) \qquad \text{where} \qquad x(a) = \sum_{i:h_i=a} w(l, i)$$

This view is fairly general. The function w specifies *where* the sample is obtained. The ξ function specifies how much it is *trusted*. The classes of adaptivity can now be defined in terms of constraints on w and ξ.

In Section 1.10.3, two constraints were placed on ξ: that ξ generate safe predictions, and that ξ converge on the naively estimated probabilities $(x(a)/y)$ at infinity. Here, a variation of Rissanen and Langdon's definition of adaptivity is used to place an additional constraint of monotonicity. δ_a is a sample containing a single instance of a. The additional constraint is

$$\forall x \in X, a \in A, \ \xi(x + \delta_a)(a) \geq \xi(x)(a)$$

---

[75] Here we take the liberty of storing real values in x which was defined in Section 1.8.2 to be a vector of integers.

If ξ is continuous, this is more simply specified as

$$\forall x \in X, a \in A, \ \frac{\partial \xi(x)(a)}{\partial x(a)} \geq 0$$

This condition in conjunction with the two of Section 1.10.3 ensures that the estimation function ξ will behave itself. The estimation function must generate safe predictions, must not decrease its estimations with increasing frequency and must converge on the naive estimations at infinity.

The four classes of adaptivity are defined by constraints on w which will now be given. For a given l, the mean position of emphasis $\overline{w}$ is defined as

$$\overline{w} = \sum_{i=1}^{l} w(l, i)i$$

For convenience, we also define

$$a = \overline{w} \qquad \text{and} \qquad b = l + 1 - \overline{w}$$

(Figure 33).

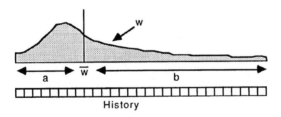

The function $w(l, i)$ can be used to describe the emphasis that an algorithm places on the instances at various positions of the history. The area under the curve is always one. In this diagram, the algorithm is weighting heavily towards the distant past. $\overline{w}$ is the mean position and a and b more concisely describe the relationship between $\overline{w}$ and the ends of the history.

Figure 33: Example adaptive weighting curve.

The values a and b give the distance of the mean $\overline{w}$ from the start and end of the history. The notation $q < K$ is used to indicate that q is bounded from above by an arbitrary finite constant. Setting $a = 0$ represents the use of no instances at all. The four classes of adaptivity can be defined in terms of the behavior of a and b as $l \rightarrow \infty$.

Non adaptive	$a = 0$	$b \rightarrow \infty$
Initially adaptive	$a < K$	$b \rightarrow \infty$
Locally adaptive	$a \rightarrow \infty$	$b < K$
Asymptotically adaptive	$a \rightarrow \infty$	$b \rightarrow \infty$

This definition is much more robust that the previous one and can classify all kinds of unusual sampling strategies. For example, a technique that assigns

$$w(l, i) = \begin{cases} odd(i) \rightarrow 2/l \\ even(i) \rightarrow 0 \end{cases}$$

would be correctly classified as asymptotically adaptive. The classification scheme does not classify models for which a and b do not converge at infinity.

The definition contains one major fault. Although it measures *where* the information is gathered, it does not measure *how much* information is gathered. A model that sets

$$w(l, i) = \begin{cases} i = \lceil l/2 \rceil \rightarrow 1 \\ i \neq \lceil l/2 \rceil \rightarrow 0 \end{cases}$$

would be classified as asymptotically adaptive even though all intuitive definitions of that class require that it accumulate an ever increasing number of instances. An improved definition must include a measure for the information content of a model. Shannon's logarithmic definition works as well here as for predictions. We define the **sampling entropy** of a model as

$$I = -\sum_{i=1}^{l} \ln w(l, i)$$

and refine our definitions for classes of adaptivity to (as $l \rightarrow \infty$)

Non adaptive			$I = 0$
Initially adaptive	$a < K$	$b \rightarrow \infty$	$I < K$
Locally adaptive	$a \rightarrow \infty$	$b < K$	$I < K$
Strangely adaptive	$a \rightarrow \infty$	$b \rightarrow \infty$	$I < K$
Asymptotically adaptive	$a \rightarrow \infty$	$b \rightarrow \infty$	$I \rightarrow \infty$

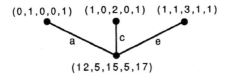

The DAFC algorithm illustrates the difficulty in classifying the adaptivity of variable-order Markov algorithms. DAFC freezes its tree when the last node has been grown, making it *structurally initially adaptive* but *contextually asymptotically adaptive*.

Figure 34: A snapshot of DAFC in execution.

This definition is satisfactory. Static models contain no information gained from the source. Initially adaptive and locally adaptive models contain a finite amount of information derived from one end of the history. Asymptotically adaptive sources contain an increasing amount of information obtained from the entire history.[76] Strangely adaptive models are the same as asymptotically adaptive models except that they contain only a finite amount of information.

3.4 Adaptivity of Previous Algorithms

In this section, the new definition of adaptivity is used to classify some of the algorithms described in Chapter 1. Algorithms that are not one pass are ignored.

One form of local adaptivity arises so often that it is worth defining a name for it. A model/algorithm is defined to be **windowed locally adaptive** if it has

$$w(l, i) = \begin{cases} i \leq l - K \to 0 \\ i > l - K \to 1/K \end{cases}$$

Knuth's windowed dynamic Huffman coding[Knuth85] and LZ77 both fall into the category of windowed locally adaptive algorithms.

In contrast, dynamic Huffman coding and LZ78 set $w(l, i) = 1/l$ and are asymptotically adaptive.

Variable order Markov algorithms are harder to classify because they employ different kinds of adaptivity at the context and structural levels. The problem is best illustrated by DAFC (Section 1.10.6.2), the simplest variable-order Markov model.

---

[76] The class of asymptotically adaptive sources could be further divided depending on the behaviour of b/a as $l \to \infty$.

DAFC allocates first-order contexts to the first 31 symbols whose cumulative frequency exceeds a fixed threshold. After this point, the tree structure (of depth one) is frozen but the contexts in the nodes of the tree continue to collect instances[77] (**Figure 34**). This means that the *structure* of the tree is initially adaptive but the *contexts* of the tree are asymptotically adaptive. Thus the DAFC algorithm is asymptotically adaptive to a Markov order of the average update depth of the tree. Because the root node is always present, the average update depth is always at least zero.

For finite memory, the same reasoning applies to PPM, DHPC and DMC. However, if there is enough memory to construct a tree whose structure encloses the source's tree, these tree algorithms all become asymptotically adaptive.

Perhaps the most important observation to be made about the relationship between adaptivity and compression techniques is that it appears that any technique can be modified to conform to any class of adaptivity. For example, LZ77 (Section 1.7.3) could be made non-adaptive by using a string buffer with a constant value. Dynamic Huffman coding (Section 1.4.4) could be made windowed locally adaptive by using Knuth's abstraction[Knuth85] to remove instances as well as add them. The satisfying conclusion is that classes of adaptivity are not bound to classes of algorithm.

3.5   A Classification of Sources

Traditionally, a source is viewed as a finite ergodic Markov source that generates an infinite string. If this were the case in practice there would be no need for anything but asymptotically adaptive models. In contrast, the success of locally adaptive techniques in practice indicates that real sources often change considerably over time. An understanding of the ways in which sources might change is necessary to approach this problem.

A changing source[78] can be viewed as a point moving in a multi-dimensional **simple-source space.**[79] For sources of a particular order

---

[77] We assume that the counters that record the frequencies of instances in each context are of infinite width.

[78] It is worth distinguishing between objects that change themselves and objects that are changed by other objects (Appendix B). Here, both sources and models are viewed as self-modifying objects.

[79] The term **source space** is reserved for the space of all moving sources.

m, simple source space consists of n^m dimensions,[80] each dimension corresponding to a probability $d(s)(a)$ of a particular symbol $a \in A$ in a particular context $s \in S_m$ of the simple source $d \in D$. Each point in simple-source space defines a finite-context, order-m Markov source (a simple source). Movement of a source through simple source space corresponds to a change in its probabilities.

In addition to moving through simple-source space, sources move through **state space**. The combination of a source's position in simple-source space $d \in D$ and state space $s \in S_m$ completely determines the next prediction that the source will generate. Because, by Assumption 7 (Section 1.10.2), the state of a model can be determined directly from the history, there are no continuity problems with the state of the changing source (as there might be for example, if the source was an arbitrary, changing finite state machine). As a simple-source's state is a function of the most recent few instances of the history, it does not determine source trajectories and will not be considered further.

Section 1.8.4 showed that the problem of data compression is to *reconstruct the source from the history*. This is accomplished by using the history to locate the source in simple-source space. Whereas fixed sources need only be *found*, moving sources must be *tracked* as they move through simple source space.

Implicit in the idea of tracking a source using a history of instances is the assumption that the history will give some indication of the source's current position. This may not be the case. If the source moves to a random position between the generation of instances tracking becomes impossible and the games theory[**Neumann44**] solution of uniform predictions must be adopted. Just as asymptotically adaptive models are most effective on a source that does not move, so are tracking models most effective on sources that move slowly.

A fundamental trade-off dominates the design of source tracking models. Whereas it is advantageous to use as many instances as possible to estimate the source's position, it is disadvantageous to use instances that are so out of date that they do not reflect the source's current position (**Figure 35**). Instances in the history reflect the position of the source *at the time they were generated*. Instances further back in the history are less informative. There is no clean solution to this problem. Practical algorithms compromise by estimating the speed of the source and using the estimation to choose a window size.

---

[80] Actually, fewer dimensions are needed because the probabilities are constrained to sum to unity. Here the redundant form is used because it is simpler. All the same concepts apply.

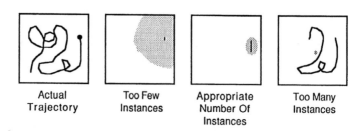

Actual	Too Few	Appropriate	Too Many
Trajectory	Instances	Number Of	Instances
		Instances	

A trade-off exists when tracking a moving source. Many instances are required to determine accurately the position of the source, but only the most recent instances are representative of its current position. In this example, the leftmost box contains the actual trajectory of the source in simple source space (ending at the dot) and the other boxes show the estimated position of the source (in grey) based on various numbers of previous instances (the black line).

Figure 35: Tracking a moving source through simple source space.

The trade-off is reminiscent of Heisenberg's uncertainty principle [**Mayfield72**](p. 134) which states that the position and momentum of a particle cannot both be known at the same time. In data compression the tracking trade-off only arises for moving sources; *all* the instances generated by a fixed source are representative of the source's current position.

The effect of the Heisenberg problem is that it is impossible to compress a source's output without making some assumptions about the source's trajectory. It seems appropriate therefore to categorize the trajectories that are likely to arise, in the hope of recognising them and forming a strategy to compress them. The following list of interesting trajectories is arranged from least to greatest entropy. Each source trajectory is depicted graphically in **Figure 36**.

Fixed: The source does not move (also called "simple").

Fuzzy Fixed: The source moves but stays close to a fixed point.

Drifting: The source moves, but never very far between each instance.

Vagrant: The source jumps to a completely new random position at irregular intervals.

Multimodal: The source jumps between a finite set of points at irregular intervals.

White Noise: The source jumps to a random position between
the generation of each instance.

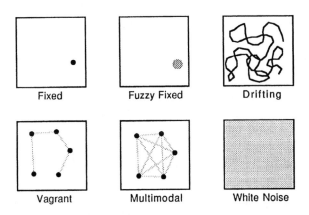

This diagram depicts graphically some interesting source trajectories.
Each box represents simple source space. A fixed source does not
move. A fuzzy source hovers around a point. A drifting source wanders
aimlessly. A vagrant source hops randomly but only occasionally. A
multimodal source occasionally switches between a finite set of points
(e.g. switches between different file types). Finally, a white noise
source moves to a new random position after each instance (the same
as vagrant but faster).

Figure 36: Some interesting source trajectories.

The list above is only intended as a rough classification and no
attempt will be made to formalize it. The adjective "fuzzy" is used
to describe a source that is imprecise in its residency at a point.

3.6   A Comparison of Kinds of Adaptivity

Having examined classes of adaptivity and classes of sources, we
are in a position to assess the effectiveness of classes of adaptivity in
compressing classes of sources.

A non-adaptive model will perform poorly unless the source is fixed
at the position assumed by the model. If the source is not fixed, the
instantaneous performance will be inversely related to the "distance"
between the model and the source.

An initially adaptive model relies not on the source being in any
particular position, but on its staying still. Initially adaptive models are
likely to perform poorly on all but fixed sources.

Asymptotically adaptive and locally adaptive models are superior to initially adaptive models. Asymptotically adaptive models use more information $(I \rightarrow \infty)$ and locally adaptive models use more recent information $(a \rightarrow \infty)$.

With non-adaptive and initially adaptive models out of the race, we consider the relative merits of locally adaptive and asymptotically adaptive models.

The advantage of asymptotically adaptive algorithms is their ability to converge on a fixed source at infinity. To achieve this, they incorporate information yielded by an ever increasing number of instances. This makes asymptotically adaptive algorithms less responsive to source movements as the length of the history increases.

The advantage of locally adaptive algorithms is their ability to track source movements quickly. They achieve this by basing predictions only on the very recent behaviour of the source. Because they use only a finite amount of information, locally adaptive algorithms are incapable of converging on a fixed source.

Thus, the choice between local adaptivity and asymptotic adaptivity should be made depending on whether convergence or responsiveness is more important. Because most real sources move around the source space, locally adaptive models are nearly always the best choice.

This concludes the theoretical part of this chapter. The remainder of this chapter contains a description of techniques for incorporating asymptotic and local adaptivity into variable-order finite-context Markov algorithms.

3.7 Mechanisms for Adaptivity

In Section 3.3, adaptivity was defined for a zero-order Markov model. The same definition can be applied to higher order models by splitting the message into different instance streams, one for each context and considering each context as a zero order model. Adaptivity can be separated into a *context component* and a *structure component*.

The theoretical discussion assumed that the entire history is available and that there is an unbounded amount of processing power available to process it. In contrast, practical models must be feasible and in particular must satisfy the following requirements.

- The model must use a fixed finite amount of memory.

- The model must process each instance in time constant with respect to the length of the history.

These requirements imply that at most only a fixed, finite portion of the history can be retained, and that models must be constructed incrementally. The only concession that we make here is to allow counting registers of infinite width. In most implementations, 32-bit registers approximate this assumption well.

3.7.1 Context Adaptivity

Context adaptivity refers to the management of instances within a particular sample. Because samples are constrained to use finite memory, they cannot store much instance ordering information. Usually they store only a frequency counter for each symbol.

Construction of an asymptotically adaptive context with $w(l, i) = 1/l$ simply involves recording the frequency of each symbol and using the frequencies to make predictions.

A windowed locally adaptive context can be obtained by maintaining a frequency for each symbol and a buffer of the K most recent instances of the **context history** which we define to be the string of instances that have occurred in a particular context. Upon the arrival of each instance, the frequency of the new instance's symbol is incremented and the frequency of the K'th most recent instance's symbol is decremented.

An alternative to maintaining a buffer is to set

$$w(l, i) = e^{-\mu i}$$

with μ set so that $\sum_{i=1}^{l} w(l, i) = 1$. Because $e^{-\mu i}/e^{-\mu(i+t)}$ is a constant for all positive i and t, w can be implemented without a history buffer by multiplying every frequency counter (after the arrival of each instance) by a constant $e^{\ln 0.5/h}$. Such an operation is called a **decay**, with h being the half life of an instance. The half life is the age (measured in instances), that an instance must be before it is half as influential as a fresh instance.

Performing n multiplications for each instance may be too expensive for practical data compression. An alternative is to perform the decay operation at regular intervals using a smaller h. This has the disadvantage of making the number of instances in samples rise and fall as a saw-tooth.

Another way to avoid frequent decay operations is to perform them only when the frequency of a symbol reaches a certain threshold τ. Although decaying will occur at irregular intervals, the technique is guaranteed to be locally adaptive because decay operations must occur at least every τn instances. This method has the practical advantage of providing an upper bound on the required width of counter registers.

3.7.2 Structural Adaptivity

The DHPC algorithm described in Chapter 2 is designed to maximize the depth of nodes used to make predictions. The algorithm does this by growing each part of the tree at a rate proportional to its use. Once all the available nodes have been added to the tree, the tree's structure is frozen. DHPC is structurally initially adaptive.

To maintain a locally or asymptotically adaptive tree structure, an algorithm must be capable of altering the structure of its tree once the tree is built. Once such an ongoing mechanism for re-organization is in place, the heuristics controlling the initial placement of nodes become much less important. For this reason, the following discussion addresses only the transformation of fully grown trees.

For the sake of simplicity, the transformations performed upon the tree will be restricted to one primitive operator called **leafmove** that moves a single leaf node from one part of the tree to another. This constraint results in no loss of generality, as the operator, repeatedly applied, is capable of transforming any tree structure to any other.

The next problem is deciding how and when *leafmove* should be applied. Because the operator transfers an anonymous (contains no information) resource from one place to another, the problem naturally splits into two separate problems: that of supply and demand.

Supply problem: Given that a leaf node must be removed from a tree, decide which leaf should be removed.

Demand problem: Given a spare node is available to add as a leaf to the tree, decide where it should be added.

Before addressing these problems, it is worth investigating metrics for gauging the worth of a leaf.

3.7.3 Metrics on Tree Structures

Many different tree structures can be constructed from a group of nodes. Of particular interest are structures that maximize the Markov order and structures that minimize the entropy. It might be that these two groups are the same.

Because trees are to be manipulated using only *leafmove*, it is important to be able to measure the worth of a leaf. Consider the worth of a leaf be (of a backwards tree) where $b \in A$ and $e \in S$. Suppose that for a fixed ergodic source we find that *at time infinity* the prediction of be is p_c (prediction of the child) and the prediction of e is p_p (prediction of the parent). If the probability of be matching the history is h_c then the absolute **entropy loss** σ yielded by the child to the tree is

$$\sigma(be) = h_c \left(- \sum_{a \in A} p_c(a) \ln p_p(a) - - \sum_{a \in A} p_c(a) \ln p_c(a) \right)$$

which reduces to

$$\sigma(be) = h_c \sum_{a \in A} p_c(a) \left(\ln p_c(a) - \ln p_p(a) \right)$$

Thus the worth of a leaf is the product of the probability of the leaf being used (h_c) multiplied by its advantage over its parent in predicting its own context. Roberts[Roberts82] and Rissanen[Rissanen83] use similar metrics.

The most powerful term in the formula above is h_c; if the node is rarely used, its entropy loss hardly matters. In the remainder of this thesis we will use frequency of use as the sole measure of the worthiness of a leaf.

3.7.4 The Supply Problem

Given that a leaf node *must* be removed from a tree, the supply problem is that of deciding which leaf to remove. Ideally we wish to remove the leaf with the lowest σ. If we settle on approximating σ by the frequency of use of a node, it would seem that the best node to remove is the node containing the least number of instances. Unfortunately, the different leaves on the tree are created at different times making this measure unfair. A better scheme is to remove the leaf with the lowest usage rate. Newly added leaves could be initialized with a slightly higher than average rate so that they are not immediately recycled.

The minimum-rate heuristic can be considered to be asymptotically converging because it will grow more and more sluggish as time goes by. The locally adaptive equivalent is to consider the rate of use of each node over a recent time period. One measure of rate is the mean time between occurrences. A crude but effective locally-adaptive lowest-rate supply heuristic is to select the *least recently used* node — the node with the lowest instantaneous rate.

3.7.5 The Demand Problem

Given that a node is available to add as a leaf to the tree, the demand problem is that of deciding where the node should be added. Constructing demand heuristics is difficult because, unlike the supply case, the candidate nodes do not already exist and the only information available about the potential leaf is in the potential parent node.

One approach is to find an upper bound on the entropy loss that a candidate leaf *could* make. Roberts[Roberts82] achieved good results by using the product of this bound and the established worth of the parent node as a metric for choosing new leaf sites.

On the basis that the best leaves are the most frequently used leaves, the DHPC demand policy of attaching leaves in proportion to the rate of use of their parent appears sensible. It is unnecessary to maintain a list of rates of usage; if a leaf is attached to the end of the matching branch with a certain fixed probability, then on average the leaves will be attached in proportion to the usage rate of each potential parent node. When an extension is made, the matching leaf is most likely to be the most likely leaf and so the most frequent child nodes will tend to be created first, resulting in a recursive effect.

3.7.6 Connecting Supply and Demand

The supply and demand systems can be organized as: independent, supply driven or demand driven. In an independent organization, the two systems operate independently. If the supply process operates faster than the demand process, the tree will soon be stripped bare (to the root) and there will be a large pool of spare nodes. If the demand system operates faster, it will soon run out of nodes created by the supply system. These problems can be avoided by placing one system in control.

In a supply driven system the supply system regularly examines the tree and remove nodes that are not performing well. These are given to the demand system which must immediately place them on the tree.

In a demand driven system, the demand system examines the tree and chooses sites for tree growth. It then requests a number of nodes from the supply system which must fulfill the order immediately.

We prefer demand systems because the demand system can easily be controlled through the flexible tree growth mechanisms.

3.8 Implementing Asymptotic Adaptivity

To be asymptotically adaptive, an algorithm must continually increase the amount of information that it stores. At the context level, this means that the frequency counters must be allowed to run to infinity. At the structure level, nodes must be added until the order of the tree matches the order of the source.

If an infinite amount of memory is available, DHPC with $m = \infty$, $z = \infty$, $\alpha > 1$ and $\beta = 0$ can be made asymptotically adaptive by increasing the credibility threshold by 1 after in^i instances have arrived for each of $i \in \mathbf{Z}[1, \infty)$. This guarantees that predictions will be made from samples whose depth and size increases to infinity.

For a model that has only finite memory[81], the situation is more complicated. For a source of order m, a tree containing at most n^m leaves is required. No other tree structure, not even a larger one, can perform better. If there is insufficient memory to build such a tree, techniques must be found for converging on a representative smaller tree.

The problem here is that at any point during compression, an asymptotically adaptive model must be prepared to move any or all of its nodes in the tree. This applies even if the source is fixed. No matter how many instances (t) the model accumulates, it is possible that they were just a statistical fluke and that at time $1000t$, the history will indicate an entirely different source position.

Unfortunately, the nodes that are being destroyed and created (moved) are where the samples are stored. When a node is moved, its instances must be destroyed or added to its parent. When this happens, information is destroyed — a happening dangerous to supposedly asymptotically adaptive models. The result is that we are faced with the ineviable task of designing an algorithm that must accumulate an infinite amount of information but which, at any time, might have to destroy all the information it currently has!

---

[81] Although we retain the infinite width registers.

Two theorems from probability theory suggest that such an algorithm could be constructed. The first theorem is that a one-dimensional random walk that is bounded from below by zero, will average infinity at time infinity. This is the same as saying that for a G/G/1/1 queue [**Kleinrock75**] with $\lambda = \mu$ (arrival rate equals service rate), the queue length will average infinite length at time infinity if the variance of the arrival and service distributions are not both zero. The second theorem is the law of large numbers.

By associating a random walk with each node in the potential tree[82] (with position k corresponding to there being k instances in the node's sample), an algorithm could be designed for which it could be shown that at infinity a given number of "optimal" nodes average an infinite number of instances. One way that such an algorithm might be built is to use a probabilistic demand heuristic and a minimum-rate supply heuristic. As time tends to infinity, the rates of each of the nodes would become less and less variable, requiring a longer run of randomness to alter their ranks.

In practice, theoretical results for asymptotic sources are of little use. Real sources often change rapidly and the extra effort involved in ensuring that an algorithm converges at infinity is not warranted. Whenever an asymptotic model is required in practice, a rough approximation usually serves well.

3.9 Implementing Local Adaptivity

Locally adaptive algorithms are much easier to construct that asymptotically adaptive algorithms because local models contain only a finite amount of information ($I < K$). We have already seen how a context can be made locally adaptive by maintaining a history buffer or by decaying. A similar technique can be used to make a structure locally adaptive.

It is possible to construct a windowed structurally locally adaptive algorithm with

$$w(l, i) = \begin{cases} i \le l - K \rightarrow 0 \\ i > l - K \rightarrow 1/K \end{cases}$$

that uses at most Km nodes of memory. At each step, the next instance is added to the tree and the K'th most recent instance is discarded from the tree. This requires only that a history buffer of $K+m$ instances be stored. Adding an instance involves growing a matching branch of depth m (if the branch is not already present). Removing an instance involves removing

---

[82] The **potential tree** contains all possible nodes and is of infinite depth.

the instance from each node on the branch that matched the history K instances before, and then removing the node if the instance removed from it was the node's last instance. Because at most K instances are present in the tree, and each instance commands a branch of depth m, at most Km nodes are required to store the tree. Typical values of K and m are $K = 5000$ and $m = 4$, for which at most 20000 nodes would be required. For a source with any measure of redundancy, only a fraction of this number would be used.

If less than Km nodes are available, a maximal length history buffer of length K or less could be maintained. Whenever the supply of nodes runs out, instances from the oldest part of the history buffer could be removed from the tree until enough nodes are freed up.

The locally adaptive algorithm described above grows a branch of length m for each new instance (if such a branch does not already exist). If memory is scarce, building long branches is wasteful because many of the deeper nodes are unlikely to collect enough instances to become credible. In many cases they will be destroyed without ever being used to make a single prediction. Reducing m is undesirable because it penalizes heavily used branches. A better solution is to use some sort of growth retardation such as DHPC's extensibility threshold. This would require the storage of an update depth for each instance in the history buffer so that the instances could be removed only from the nodes to which they were originally added.

It should be noted that it is possible to incorporate structural local adaptivity without directly incorporating context local adaptivity. This can be done by using an LRU supply system. This scheme yields an effect identical to the variable length buffer scheme (with an infinite maximum length) except that it only removes instances when a node is moved. This allows structural and context adaptivity to be separated.

3.10 Summary

In this chapter, the definition of adaptivity has been refined by constraining the estimation function ξ and by using a weight function w to define four classes of adaptivity: non-adaptive, initially adaptive, locally adaptive and asymptotically adaptive. This definition of adaptivity focuses on the emphasis that a technique places on various parts of the history. The variability of real world data is modelled by considering it to have been generated by a source moving through a Markov simple-source space. Such moving sources can be compressed by tracking them through the source space using the instances they generate as a trail. A trade-off arises because the usefulness of an instance in determining the current position of the source decreases with the instance's age. Whether it is better to use an asymptotically adaptive model or a locally adaptive model depends on whether the source is fixed or moving.

Markov models can be modified to be adaptive in any of the four ways by manipulating the instances within each context (contextual adaptivity) and by manipulating the structure of the tree (structural adaptivity). Locally adaptive algorithms are easy to construct because they require only a finite amount of information. Asymptotically adaptive algorithms present greater difficulties, but could be constructed (within the constraints of the memory) using probabilistic techniques. In practice, sources move often enough to render such fine tuning irrelevant.

The next chapter describes an algorithm that uses the contextual and structural mechanisms described in this chapter to implement local and asymptotic adaptivity.

Chapter 4

An Experimental
Adaptive Algorithm

4.1 Introduction

In this chapter an algorithm is presented that incorporates many of
the mechanisms for adaptivity described in Chapter 3. The algorithm
began as DHPC and developed incrementally. Originally, the algorithm
was to be used to investigate the performance of variants of DHPC and
PPM. However, as the algorithm developed, it became clear that the
algorithm's flexibility and integration of diverse, interacting features was
of interest in its own right. In this chapter, the algorithm, called the
SAKDC[83] algorithm, is discussed in detail and the results of experiments
that explore its parameter space are presented. The exploration of
SAKDC's parameter space not only lends experimental support to the
theory presented in Chapter 3, but provides guidelines for practitioners
working with Markov algorithms.

It should be stressed from the start that SAKDC is an experimental
algorithm in which flexibility and reliability have taken precedence over
efficiency. In particular, the interaction and in some cases the very
existence of parameters has prevented many optimizations from being
made. In a production compressor, these parameters would be fixed,
and many optimizations could be made. For example, if the depth were
fixed at 3, loops could be unrolled.

---

[83] Swiss Army Knife Data Compression.

Paper	Value	Description
Williams88	β/α	Compression
Bahl74	α/β	Compression
Jakobsson82	α/β	Compression gain
Pratt80	α/β	Compression ratio
Teuhola78	α/β	Compression gain
Turner75	α/β	Compression ratio
Bentley86	$8(\beta/\alpha)$	Bits per character
Brent87	$8(\beta/\alpha)$	Compression ratio
Cleary84	$8(\beta/\alpha)$	Bits per character
Cormack85	$8(\beta/\alpha)$	Bits per character
Moffat88	$8(\beta/\alpha)$	Compression
Roberts82	$8(\beta/\alpha)$	Entropy
Cormack85	$100(\beta/\alpha)$	Compression ratio
Cormack87	$100(\beta/\alpha)$	Compression factor
Gottlieb75	$100(\beta/\alpha)$	Compression ratio
Mcintyre85	$100(\beta/\alpha)$	Not named
Moffat87	$100(\beta/\alpha)$	Compression ratio
Cooper78	$100(1 - \beta/\alpha)$	Compression
Cooper82	$100(1 - \beta/\alpha)$	Compression
Katajainen86	$100(1 - \beta/\alpha)$	Compression gain
Mayne75	$100(1 - \beta/\alpha)$	Space saving
Mccarthy73	$100(1 - \beta/\alpha)$	Storage saving
Raita87	$100(1 - \beta/\alpha)$	Percentage saving
Reghbati81	$100(1 - \beta/\alpha)$	Compression
Severance83	$100(1 - \beta/\alpha)$	Compression
Tischer87	$100(1 - \beta/\alpha)$	Compression ratio

The method of reporting compression performance varies greatly from researcher to researcher. This table lists the measures employed in an informal selection of papers. α is the size of the uncompressed message and β is the size of the compressed message. The description column lists the descriptions used by the different authors.

Table 4: Compression performance measures.

1.16   Error Correction, Data Compression and Cryptography

Throughout the development of electronic communication systems there has been a strong link between error correction, data compression

4.2   Overview of the SAKDC Algorithm

SAKDC is an extension of DHPC and contains DHPC as a special case. Like DHPC, SAKDC uses a backwards tree, has a depth limit m, a maximum number of nodes z, and a credibility threshold β. DHPC and SAKDC differ in the collection and disposal of instances and the management of nodes.

SAKDC uses a demand-driven supply/demand system. The demand system uses an extensibility threshold to determine if a node is fit for growth and then appends matching leaves with a specified probability. Two sets of growth parameters are used, one set for trees containing z node and one set for trees with less than z nodes. The supply system is LRU (Least Recently Used). Instances can be added to the entire matching branch or just to some of the nodes on the branch. Samples are decayed when a sample reaches a threshold. A windowed local adaptivity option allows instances to be forgotten after K more instances have arrived. Shortcut pointers are used to speed up tree traversal, with an aging technique used to detect pointers that have been invalidated by node movement.

SAKDC's strengths lie in its flexibility and its sophisticated adaptivity options that allow it to run indefinitely without becoming initially adaptive.

The algorithm will be described by listing its parameters and then discussing specific implementation problems and their solutions.

4.3   Parameters of the SAKDC Algorithm

This section describes SAKDC's functionality by specifying the parameters that control its behaviour. In this and following sections, the author has opted for precision over conciseness in the hope of avoiding the ambiguity present in so much of the literature.

Some common structures arise repeatedly in the parameters. A **threshold** is a tuple (*kind,thresh*) that is used to classify all samples as either under or over the threshold. *thresh* is a positive integer threshold. *kind* is an enumerated type having one of the two values *sum* or *max*. A sample is defined to be over a threshold t iff *t.kind=sum* and the sample contains at least *t.thresh* instances or if *t.kind=max* and the frequency of the symbol whose frequency is maximum is at least *t.thresh*.

The SAKDC parameters are collected in **Table 7** for easy reference. Parameters of the form *P.active* determine whether feature P is turned

Parameter	Description
maxdepth	Maximum depth of the tree.
maxnodes	Maximum number of nodes in the tree.
grow.active	Does growth occur in a growing tree?
grow.threshold	Extensibility threshold for a growing tree.
grow.probext	Probability of extension for a growing tree.
move.active	Does growth occur for a moving tree?
move.threshold	Extensibility threshold for a moving tree.
move.probext	Probability of extension for a moving tree.
lruparent	LRU discipline.
phoenix.active	Will tree be destroyed when memory runs out?
phoenix.ashes	Number of past instances used to reconstruct tree.
decay.active	Will decaying take place?
decay.threshold	Threshold above which decaying takes place.
decay.factor	Decaying multiplies frequencies in sample by this.
decay.rounding	Are frequencies rounded when decayed?
decay.residue	Does decaying keep positive frequencies positive?
local.active	Windowed local adaptivity to be used?
local.period	Time after which instances are removed from tree.
deeponly	Determines how much of matching branch is updated.
addback	Parents inherit instances of doomed child nodes?
shortcuts	Use shortcuts to reduce execution time?
estim.threshold	Nodes are credible if their sample is over this.
estim.merge	Determines the method of merging nodes.
estim.method	Estimation method.
estim.lambda	Confidence parameter on estimation technique.

Table 7: Summary of SAKDC parameters.

on or off. If *P.active=false*, the remaining parameters of the form *P.** are meaningless.

The next few sections describe the parameters of SAKDC, giving an overview of the effect of each parameter, but deferring to later sections detailed discussion of the parameter's implementation and interaction with other parameters.

4.3.1   Tree Growth

Ten parameter groups control the growth of the tree.

maxdepth: $Z[0, \infty)$. This is the maximum allowable depth of any node in the tree.[84] The *maxdepth* parameter corresponds to m in DHPC and will often be referred to as m.

---

[84] Reminder: The root node is at depth zero (Section 1.2).

maxnodes: $Z[1, \infty)$. This is the maximum number of nodes allowed in the tree. If the tree contains less than the maximum number of nodes, it is called a "growing" tree, otherwise it is called a "moving" tree. If windowed local adaptivity (Section 4.3.4) is turned on, it is possible that the tree will alternate between growing and moving. The *maxnodes* parameter corresponds to z in DHPC and will often be referred to as z.

grow, move: *record.* The remainder of the tree growth parameters are divided into two groups of identical structure called *move* and *grow*. During execution, the *grow* group is used if the tree contains less than *maxnodes* (i.e. z) nodes, otherwise the *move* group is used. In a growing tree, the addition of a node requires the creation of a node. In a moving tree, addition of a node requires the removal of a different node from the tree.

grow.active, move.active: *boolean.* If this parameter is *false*, no tree growth occurs. If this parameter is *true*, the remainder of the *grow* or *move* parameters determine the conditions under which a node is added to the tree.

Whenever an instance arrives, it is added to some nodes on the matching branch. Once this is done, one or more nodes can be added to the end of the matching branch. A node will be added only if *regime.active=true* and the candidate-parent node's sample exceeds an extensibility threshold. If both these conditions hold, a new matching node is attached with a certain probability (the probability of extension). If a node is attached, the process repeats with the new node having to meet the requirements afresh.

grow.threshold, move.threshold: *threshold.* A node is defined to be extensible if it exceeds this threshold.

grow.probext, move.probext: $R[0, 1]$. If the potential parent node is extensible, it then has to pass a probability criterion. The new node is only created with a certain probability, the **probability of extension**. A deterministic random number generator is used to implement this constraint.[85]

The semantics of the tree growth parameters are summarized by **Figure 37.**

---

[85] The random number generator must be deterministic so that its behaviour can be reproduced by the decompressor (Section 1.11.6).

{*current* is the tip of the matching branch}
loop
    regime : (*move*, *grow*);
    exit if *depth*(*current*)=*maxdepth*;
    if *nodes_in_tree*=*maxnodes* **then**
        regime ← *move*;
    else
        regime ← *grow*;
    end if;
    exit if not *regime.active*;
    exit if not *is_over_threshold*(*current.x*,*regime.threshold*);
    exit if not *random*(0,1) <*regime.probext*;
    newnode : *node*;
    new(*newnode*);
    current.child[*hist*[*depth*(*current*)+1]] ← *newnode*;
    current ← *newnode*;
end loop;

Unlike the DHPC algorithm (whose tree grows at most one node per instance), the SAKDC algorithm can add many matching nodes during a single update. After the matching branch has been updated, nodes are added iteratively by the **loop** until one of the exit conditions becomes *true*. At the start of each iteration, one of the two growth regimes (*move* or *grow*) is selected depending on whether there are *maxnodes* in the tree. Then, if all the conditions are met, a new matching node is added to the end of the branch. The node is then updated (not shown here) with the new instance, and a new iteration commences starting from the newly created leaf node.

Figure 37: Summary of growth parameters.

4.3.2 The Supply System

The parameters described in the previous section determine the conditions under which a new node will be added to the tree. The new nodes are obtained from the supply system. The SAKDC supply system is controlled by the following three parameters.

lruparent: (*youngest*,*same*,*oldest*). The default supply system in SAKDC is to remove the least recently used node (set *lruparent*←*same*). Unfortunately, maintaining this information is computationally expensive (Section 4.7), and it is worth investigating slightly dirtier methods. The *lruparent* parameter determines where in the LRU list to insert nodes whose only child has just been removed. If *youngest*, the node is placed at the head of the LRU list (position least likely to be recycled). If *oldest*, the node is placed at the tail of the LRU list (position most likely to be recycled). If *same*, the node is placed in a position consistent with the time it was last used (i.e. LRU order is maintained). See Section 4.7 for more on this parameter.

phoenix.active: *boolean.* If *true*, the compressor's tree is completely destroyed (and its nodes placed in a pool for recycling) whenever the supply of nodes runs out. In order to prevent the new tree from initially performing poorly, the new tree is primed with the most recent *phoneix.ashes* of the history. The *phoenix* parameter was included so as to allow other authors' compression algorithms to be selected (in particular Moffat's PPMC' algorithm[Moffat88]). Many authors advocate destroying and rebuilding data structures when memory runs out.

phoenix.ashes: $Z[0, \infty)$. This parameter specifies the number of instances used to rebuild the tree after its destruction.

4.3.3   Decaying

The next group of parameters determines how the instances within each node will be decayed. The motivation for decaying is given in Section 3.7.1.

decay.active: *boolean.* If this parameter is *false*, decaying does not take place; if *true*, it takes place whenever a sample exceeds *decay.threshold*.

decay.threshold: *threshold.* A decay operation takes place whenever a sample exceeds this threshold.

decay.factor: $R[0, 1]$. A decay operation has the effect of scaling (multiplying) each frequency in the sample by this parameter.

decay.rounding: *boolean.* If *true*, frequencies are rounded after being scaled. If *false*, the frequencies are truncated.

decay.residue: *boolean.* If *true*, frequencies that were truncated or rounded down to zero but were previously positive are set to one. If *false*, no action is taken.

Whenever an instance is added to a sample, a check is performed to see if the sample should be decayed. The semantics of the decay operation are given in **Figure 38**.

```
inc x(newinstance);
if decay.activity then
    if over_threshold(x,decay.threshold) then
        for a in A loop
            oldxa ← x(a);
            if decay.rounding then
                x(a) ← ⌊ x(a) × decay.factor + 0.5 ⌋;
            else
                x(a) ← ⌊ x(a) × decay.factor ⌋;
            end if;
            if decay.residue and oldxa> 0 and x(a)= 0 then
                x(a)← 1;
            end if;
        end loop;
    end if;
end if;
```

> This code describes the semantics of the decay parameters of the SAKDC algorithm. Decaying is used to introduce local adaptivity at the context level. Decaying also places an upper bound on the frequency counts in a sample. Whenever an instance is added to a sample, the code above is executed. If the sample exceeds the decay threshold *decay.threshold*, each frequency in the sample is multiplied by a decay factor *decay.factor*. Rounding and residual parameters take care of details.

Figure 38: Semantics of decaying.

4.3.4 Windowed Local Adaptivity

The parameters described so far allow local structural adaptivity and local sample adaptivity. Each of these mechanisms employs a roughly negative exponential w function (Section 3.3). Another important form of local adaptivity is windowed local adaptivity (Section 3.4). A windowed locally adaptive model *completely* forgets an instance after a fixed interval. The *local* parameters allow windowed local adaptivity to be specified.

local.active: *boolean.* *true* iff windowed local adaptivity is to be employed.

local.period: $Z[1, \infty)$. If *local.active=true*, each instance is removed from the tree *local.period* instances after it arrives. If an instance removed from a node was the only instance in the node, the node is removed from the tree for recycling.[86]

4.3.5 Instance Management

DHPC adds each new instance to every node in the matching branch. SAKDC allows more complicated updating.

deeponly: (*whole,credible,symbol*). This parameter specifies the policy used to determine which nodes on the matching branch are updated (have the new instance *a* added to their sample). If *whole*, all nodes are updated. If *credible*, only nodes at least as deep as the deepest credible node are updated. If *symbol*, only nodes at least as deep as the deepest node containing an instance of the symbol of *a* are updated.

addback: *boolean*. If *false*, a node's instances are destroyed when the node is moved. If *true*, instances that were not originally added to the parent of the node about to be moved, are added to the parent of the node about to be moved.

4.3.6 Improving Efficiency

Execution of DHPC involves traversing the matching branch once for each instance processed. This means that the processing of each instance takes $O(m)$ time. To avoid this traversal, shortcut pointers (Section 1.12.2) can be associated with each symbol in each sample in the tree. The pointer associated with symbol *a* in node *x* points to a node that matches *xa*.[87] The algorithm endeavors to make these pointers point to as deep a matching node as possible. When a new instance arrives, a shortcut pointer is used to jump deep into the next matching branch. This saves traversal time, especially if *deeponly=true*.

shortcuts: *boolean*. If *false*, the tree is traversed from the root at each step. If *true*, shortcut pointers are used to jump deep into the matching branch upon the arrival of each new instance. This parameter has no effect on the functionality of the algorithm, only on its efficiency.

---

[86] This can cause the status of the tree to change from *moving* to *growing*.

[87] This is a variant of the *AddRight* operator of Section 1.12.2 which mapped *a* and *x* onto *xa* only. Here shortcut pointers can point to any tail string of *xa*.

4.3.7 Estimation

SAKDC has four estimation parameters. Estimation is oriented around a pivot node on the matching branch. The **pivot node** is the deepest credible node on the matching branch.

estim.threshold: $Z[1, \infty)$. A node is termed "credible" if its sample is over this threshold.

estim.merge: (*DHPC,LAZY,PPM*). The merge[88] parameter determines whether DHPC, LAZY or PPM node merging is to be used. In DHPC, the sample of the pivot node is used to make the prediction. In PPM, the samples of the pivot node and its ancestors are blended together to make the prediction (Section 1.10.6.6 and **Figure 15**). LAZY is the same as PPM except that it does not perform "exclusions" (**Figure 39**).

All three merging techniques use the estimation function $\xi(x)$ specified by the remaining *estim* parameters.

estim.method: (*linear, nonlinear, linear_moffat, nonlinear_moffat*). This parameter determines the estimation formula used to estimate the probabilities of symbols. The formula can be linear (Section 1.10.3.1), non-linear (Section 1.10.3.2), linear Moffat or non-linear Moffat (Section 1.10.3.3). In the case of PPM merging (*estim.merge=PPM*), the sum of the probabilities allocated to zero frequency symbols by the estimation technique is used as the escape probability[Cleary84].

estim.lambda: $R(0, \infty)$ is the confidence parameter of the estimation method chosen (Section 1.10.3).

4.4 Representation of Predictions

Previous sections describe the parameters and functionality of the SAKDC algorithm. This and future sections describe the problems that were encountered during the algorithm's implementation, and the solutions that were adopted. This section describes the representation used for predictions. The terms "prediction" and "sample" will be used interchangeably.

Section 1.11 showed that choosing a data structure for predictions (samples) is non trivial. In the case of SAKDC, the prediction abstraction was further complicated by the following requirements.

---

[88] This parameter should probably be called *estim.blend* after other authors' nomenclature (e.g. [Bell89]).

```
function LAZYest (xs : samples) return prediction;
    pavail : real ← 1.0;
    notdone : constant real ← ∞;
    pred : array(symbol) of real ← (others → notdone);
begin LAZYest
    for order in reverse −1 . . . m loop
        x : sample ← xs(order);
        zerosum : real ← 0.0;
        for a in A loop
            if x(a)=0 then
                zerosum←zerosum+ξ(x)(a);
            end if;
            if pred[a]=notdone and x(a)>0 then
                pred[a]←pavail×ξ(x)(a);
            end if;
        end for;
        pavail←pavail×zerosum;
    end for;
    return pred;
end LAZYest;
```

The LAZY estimation algorithm lies half way between the DHPC and PPM estimation algorithms in computational expense and performance. Starting at the tip of the matching branch, the LAZY algorithm allocates probability in the same manner as the PPM algorithm — the available probability is divided among the symbols of positive frequency leaving a little left over (the "escape" probability) for the symbols of zero frequency. As in the PPM algorithm, this escape probability is divided according to the sample of the parent node. However, the LAZY algorithm does not exclude symbols already seen at higher levels. The LAZY estimation algorithm can be compared with the PPM estimation algorithm (**Figure 15**).

Figure 39: The LAZY estimation algorithm.

• There must be a *decay* operation.

• A shortcut pointer must be associated with each symbol with a positive count.

• The *addback* operation requires that two frequency counts be stored for each symbol whose count is positive: one for the number of instances and another for the number of instances also received by the parent node. An operation to add all the instances of one sample to another sample is required.

- Because nodes can be destroyed, so can predictions. A destroy routine is required if a dynamic data structure is used.

- Frequencies must be decremented as well as incremented, so as to allow the local adaptivity mechanism to remove instances from samples.

- The abstraction must be able to quickly provide the number of instances in a sample and the maximum frequency in the sample.

- Incremental PPM merging must be supported.

Many of these operations can be expensive $(O(n))$. Luckily, it can be shown that SAKDC uses the operations in a manner that guarantees a low *amortized* cost. Proofs of this can be based upon the fact that at most $m + 1$ instances are added to the tree upon the arrival of each instance.

For example, if predictions are represented by a sparse data structure, the time taken to destroy a prediction is bounded by the number of symbols in the prediction whose frequency is positive. This in turn is bounded by the number of instances in the prediction. As at most $m + 1$ instances are added to the tree each time a new instance arrives, the amortized cost of all the calls of the *destroy* operator in DHPC can be at most $m + 1$ calls per instance. Another example of such a proof is given in Section 4.9.

The greatest difficulty is the simultaneous requirement for: an operator to increment a frequency, an operator to decrement a frequency, and an operator to return the maximum frequency. A constant time solution seems impossible. Although a logarithmic solution is possible, our implementation simply stores a variable for the maximum frequency and updates it by performing an $O(n)$ search whenever a decrement operation takes place.

The data structure actually used was an unbalanced binary tree. Unbalanced trees are less likely to become as slow as a list, but still avoid the implementation complexity of balancing. Each frequency counter was four bytes wide, allowing large files to be compressed with decaying turned off.

The final record structure for each node in the prediction tree consisted of a symbol name, the cumulative frequency of its subtree, the number of instances of the symbol received by the parent, a shortcut pointer and left and right pointers.

The final prediction abstraction supported coding (the generation of output bytes) only for DHPC merging with linear estimation and $\lambda = 1$. In order to map the estimated probability of a symbol onto an integer range suitable for use by the arithmetic code, the numerator and denominator of the linear estimation formula in Section 1.10.3.1 were multiplied by n yielding the following formula.

$$\xi(x)(a) = \frac{nx(a) + 1}{ny + n}$$

This led to problems with register widths. For $n = 256$ and files of lengths approaching 2^{24} symbols, $ny+n$ can grow larger than 32 bits. For smaller files it can get dangerously close to 24 bits. A six-byte coder solved this problem but proved inefficient because it had to be implemented using double-precision floating point arithmetic. Far better implementations are possible (Section 1.11) but were not pursued, as the main focus of this work is on modelling, not coding.

4.5   Representation of the History Buffer

The history buffer is a data structure that stores the most recent m instances received from the source. It should not be confused with the "history" which is the name for the string consisting of *all* the instances received from the source. The history buffer consists of m slots each of which contains an instance. The slots are numbered $\mathbf{Z}[1, m]$ with slot 1 holding the most recently received instance (the youngest) and slot m holding the least recently received instance (the oldest) (**Figure 40**).

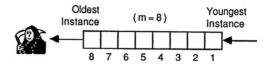

A direct implementation of a history buffer uses an array as shown. Whenever a new instance arrives, the entire array is shifted. This requires $O(m)$ assignments. However, if m is small, this may be the most efficient implementation.

Figure 40: Direct implementation of a history buffer.

When a new instance arrives, it is used in conjunction with the history to update the Markov tree. This done, the instances in the history buffer are shifted one slot. The instance in slot i is moved to slot $i + 1$. The instance in slot m is discarded and the newly arrived instance is placed in slot 1. History buffers provide the same functionality as fixed-length queues but also allow random read access to any of their elements.

A history buffer is most simply implemented as an array, with the shift operation being performed by a loop (**Figure 40**). If m is large, the shifting operation can become expensive, warranting the use of a cyclic bounded buffer which eliminates the shift at the cost of modulo arithmetic at every access (**Figure 41**). In practice, m is usually so small (≈ 4) that a cyclic bounded buffer is less efficient. In a production compressor, for a small fixed m, the shifting operation of a direct array representation could be hard coded using direct assignments or move instructions.

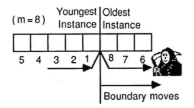

A cyclic implementation of a history buffer uses an array but does not lock the ends of the buffer to the ends of the array. Instead, the position of the ends of the buffer moves through the array. The $O(m)$ shifting operation is eliminated, but a **mod** operation is necessary on each access. This representation is appropriate only for large m.

Figure 41: Cyclic array implementation of a history buffer.

Our implementation of SAKDC used a cyclic bounded buffer. This was a consequence of having originally written the compressor with larger m in mind.

4.6 Representation of the Tree

The SAKDC algorithm uses a backwards tree supporting the following set of mappings: {*AddLeft, StripLeft, AddRight*} (Section 1.12.2).

The *StripLeft* mapping is implemented by including a parent pointer in each node. Although parent pointers are strictly unnecessary, they can greatly increase efficiency. Parent pointers allow a branch to be traversed from leaf to root at the cost of only one pointer access per arc.

This is very much faster than a root to leaf traversal which costs (say) one binary tree search per arc. The advantages of parent pointers are discussed further in Section 4.10.

The *AddRight* mapping is implemented by associating a pointer with each symbol in the prediction data structure (Section 4.4).

The *AddLeft* mapping maps nodes in the backwards tree to their child nodes (**Figure 42**). Each node can be required to store up to n pointers to child nodes. As the tree is likely to be sparsely branched, use of an array of n pointers would be wasteful.[89] **Figure 43** shows an array implementation of a tree structure that wastes most of its array space on null pointers. A sparse data structure is more profitably employed.

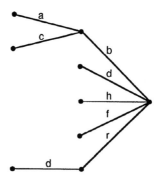

Figure 42: A branching tree structure.

Choosing a furcation representation is easier than choosing a prediction representation because the furcation representation does not need to comply with the needs of the coder. A furcation representation merely has to provide an efficient sparse mapping from symbols to pointers, allowing *insert*, *delete* and *lookup* operations. We choose unbalanced binary trees (**Figure 44**).

Figure 44 shows that there is a one to one correspondence between the node in each furcation binary tree and the Markov tree node that it points to. It seems sensible therefore to incorporate the furcation binary tree into the nodes of the main tree structure (**Figure 45**). Each node in the main tree stores a left and right pointer as part of its parent node's child tree and a pointer to a tree of its own children. This small modification contributes significantly to efficiency by eliminating a whole class of heap object. It is tempting to try to apply the same

---

[89] The wastefulness of the array is bounded by a constant factor n.

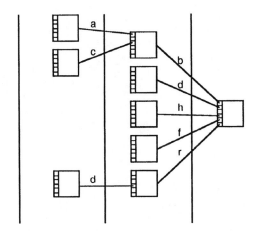

Because most trees are sparse, an array implementation of furcations
is wasteful of space; most pointers point nowhere.

Figure 43: Branching using an array.

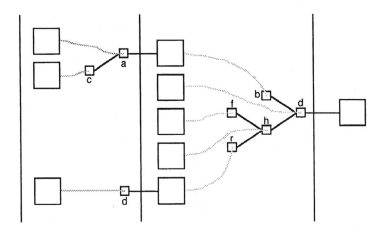

The array can be replaced by a binary tree whose sparsity conserves
memory. However, this also introduces a new heap object.

Figure 44: Branching using an explicit branch tree.

trick to the prediction binary trees whose nodes store symbol frequencies
and shortcut pointers. However, the trick is not applicable because in
SAKDC many shortcut pointers can point to the same node.

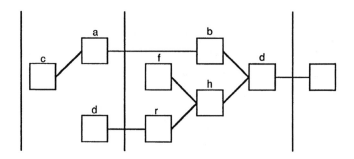

The one to one relationship between the explicit furcation tree of
Figure 44 and the nodes it points to means that the furcation tree
can be incorporated into the nodes themselves.

Figure 45: Branching using a branch tree built into the main tree.

4.7 Maintaining LRU Information

SAKDC uses demand driven node migration and the LRU supply
system suggested in Section 3.7.4. The demand system is fairly simple;
at the end of each update phase, zero or more nodes are added to
the matching branch (Section 4.3.1). The supply system is much more
complicated because it involves the maintenance of LRU information.

The LRU supply heuristic described in Section 3.7.4 ranks the
leaves according to their instantaneous frequency, this being inversely
proportional to the time since the leaf was last used. A node is *used*
whenever an instance is added to it during an *update* operation.

Unfortunately, the task of keeping track of the oldest leaf is non-
trivial. Usually, LRU algorithms from many areas maintain a doubly
linked list and move each element to the front of the list whenever the
element is accessed. This would be ideal for this application were it
not for the fact that in the SAKDC tree, leaves regularly change into
non-leaves and non-leaves into leaves.

4.7.1 The Two-Colour LRU Maintenance Problem

The SAKDC LRU maintenance problem can be expressed in abstract
terms as follows. An abstract data type T must maintain the LRU

relation of a set of r distinct things. Each thing is coloured either red or green.[90] The following operations are defined on T.

init(T) — Initializes the set of things to the empty set.

insert(T,t) — Places thing t under the control of T.

use(T,t) — Records that the thing t has just been "used".

paint(T,t) — Changes the colour of thing t.

grab(T) — Returns the least recently used green thing.

The above operations are applied one at a time to arbitrary things. The *init* and *insert* operations are included only for completeness and the discussion will concentrate on the other three operations. The problem at hand is to find an efficient implementation for this ADT.

The simplest solution is to use a linked list of things, with the thing at the head of the list being the most recently used thing and the thing at the tail of the list being the least recently used thing. The *use* operation moves a thing to the head of the list. For this representation, *use* operations can be performed in constant time as can *paint* operations. The difficulty arises when a *grab* operation must be performed. To obtain the least recently used green thing, a search must be made from the tail of the list forwards until a green thing is found. This is expensive in both theory and practice.[91]

To avoid these long linear searches, one might consider threading the list with a list of the green things. This would allow the *grab* operation to be performed in constant time. Unfortunately, this has the effect of forcing the *paint* operation to perform a linear search whenever a red thing is changed into a green thing. Similar trade offs arise for a variety of other list organizations.

One interesting organization is to store the time of the most recent use of each thing (in each thing) and maintain only a list of green things. As before, this yields a constant time *use*, a constant time *grab* and a linear *paint*. The appeal of this organization is that it strips the problem to its essentials. All that need be found is a fast way of inserting a just-painted green thing into the correct position of the green thing list. As

---

[90] Red and green are the colours traditionally used in queueing theory in such situations. In this case red corresponds to the non-leaves and green to the leaves. Think of it as a redwood tree with green leaves.

[91] If shortcut pointers are used (Section 4.10), all the rarely used non-leaf (red) nodes cluster at the tail of the list resulting in extremely long searches to find the LRU green node (typically 500 nodes in a 20000 node list).

there seemed to be no constant time solution to this problem, a heap structure was adopted. The heap structure is described in detail in the next section.

It is tempting at this stage to avoid the heap structure by inserting newly painted green things at the head or tail of the green list rather than at their correct position. Both these ideas (along with the heap) were implemented as options of the *lruparent* parameter.

The policy of inserting newly-painted green nodes at the head of the list (*youngest*) means that the minimum time needed to destroy a rarely accessed branch is increased. Each time the leaf of the branch is removed, the parent will be placed at the head of the list. This protects it from destruction for at least another z instances. Thus this policy ensures that it will always take at least zd instances for a branch of depth d to be destroyed.

Inserting newly-painted green nodes at the tail of the list (*oldest*) would place the nodes in immediate danger of being recycled. This would accelerate the removal of dead wood. It would also mean that rarely used child nodes might endanger their heavily-used parent nodes when they are recycled. In fact, the parent node is only placed in danger if its *last* child node is removed. Because the demand system tends to grow child nodes from heavily used parent nodes, it is unlikely that the *only* child of a heavily used node will be lightly used.

4.7.2   A Heap Implementation

This section describes a heap implementation that yields *use* and *paint* operations logarithmic in the number of leaves and a constant time *grab*.

A monotonically increasing counter representing the time in instances (e.g. $|h|$) is maintained. The counter is incremented each time a *use* occurs. Each thing has an "age" which is the time it was last used. A heap of green things is maintained by age with the oldest thing (lowest age value) at the top of the heap.

Using these structures, the operations are implemented as follows:

use(T,t) — Thing t's age is updated to the current time. If t is red, no action is taken. If t is green, t (already in the heap) is sifted into a position in the heap consistent with its new age. This operation takes time logarithmic in $|T|$.

paint(T,t) — If t is red, it is painted green and inserted into a position in the heap consistent with its age. If t is green, it is painted red and removed from the heap. This operation takes time logarithmic in $|T|$.

grab(T) — The least recently used green thing is always at the top of the heap and can be located in constant time.

4.7.3 Implementing a Heap using Dynamic Data Structures

Traditionally, heaps are implemented using arrays. However, as arrays cannot usually be extended incrementally, it was desirable to avoid them in favour of dynamic data structures. The result was a heap implementation technique that to the author's knowledge is original.[92]

The heap resides entirely in the dynamic storage area. The heap is structured as shown in **Figure 46** and has the following properties.

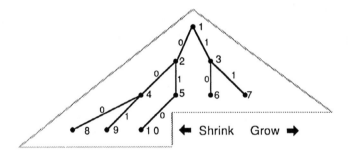

By carefully numbering heap positions, a heap can grow and shrink using dynamic memory management. The binary representation of the number of each slot (node) in the heap describes its path from the root to the slot. The heap always takes the form of a solid binary tree with a partially (left to right) filled bottom layer. Nodes are added and deleted at the bottom of the heap.

Figure 46: The structure of the heap.

[92] This solution was arrived at with the help of Barry Dwyer of the Department of Computer Science, The University of Adelaide (November 1987).

• Each node in the heap is numbered according to the path taken in getting from the root node to the node. Starting with a 1, binary digits are added to the least significant (right) side of the number until the node is reached. The root node has the number 1. Node 3 has the number 11. This organization means that the nodes in the heap are numbered consecutively along each level.

• A tree of r nodes consists exactly of those nodes numbered $Z[1, r]$. This means that the heap is always well balanced. New nodes are added at the $r + 1$'th position. An old node can be deleted by swapping it with the node at the r'th position and then removing it.

• The age of each node of the heap is less than or equal to that of its child nodes.

Insertions and deletions can be implemented as follows.

Insertion: To insert a node into the heap, the node is attached at position $r + 1$ and sifted up until its parent node is as least as old as the new node.

Deletion: To delete a target node, currently numbered k, from a heap of r nodes, the target node is first swapped with the r'th node in the heap (to be called the "refugee" node). The target node is then removed from the heap, leaving $r - 1$ nodes. Removing the target node is easy because, having become the r'th node, the target node has no child nodes. Following this, the refugee node is sifted into a consistent position in the heap.

These operations require that each node store a pointer to its heap parent. A sift operation takes time logarithmic in $|T|$.[93] The advantage of this heap organization is that it allows indefinite growth using non-contiguous, dynamically allocated memory.

4.7.4 Boundary Problems

Two small problems remain for the supply system.

The first problem is that of ensuring that the demand system is never supplied with the node that it is about to build upon (the build node). A simple solution is for the demand system to "use" the build node before requesting a leaf from the supply system. This moves the build node to

---

[93] Because the heap is balanced, this is an exact upper bound, not an amortized upper bound.

the end of the LRU list which prevents it from being chosen next (so long as there are at least two leaves).

The second problem occurs if the build node is the only leaf in the tree. This can occur if $z \in \mathbf{Z}[1, m]$, and can be avoided by preventing growth under those circumstances or by using more than m nodes.

4.8 Deep Updating

The *deeponly* parameter determines how much of the matching branch is updated by each arriving instance. When *deeponly=whole*, each arriving instance is added to every node in the matching branch. When *deeponly=credible*, each arriving instance is added only to the deepest credible node and its matching descendents. When *deeponly=symbol*, each arriving instance is added only to the deepest node that already contains an instance of the arriving instance's symbol, and that node's matching descendents (**Figure 47**). The *whole* option is used by PPM and DHPC. The *credible* option was invented by the author of this thesis. The *symbol* option was invented by Moffat[Moffat88]. In this thesis, the *deeponly* parameter is sometimes referred to as being "turned on" (*deeponly≠whole*) or "turned off" (*deeponly=whole*). "Deeponly updating" refers to updating with *deeponly≠whole*.

Under ordinary updating (*deeponly=whole*, as in DHPC), a non-leaf node accumulates an instance each time its context occurs, but is used to make predictions only when none of its child nodes are able to (i.e. when there is no credible matching child node). This means that each node's sample is collected over a superset of the conditions under which it can be required to make a prediction. The values *credible* and *symbol* of the *deeponly* parameter cause instances to be added to a node only under the conditions in which the node's sample could be used to make a prediction.

The difference between *deeponly=whole* updating and *deeponly=credible* updating is shown in **Figure 48**. For simplicity's sake, the diagram assumes a credibility threshold of one.

If *deeponly=credible* and the extensibility threshold is less than the credibility threshold, there can be a twilight period for each non-leaf node during which it accumulates instances in contexts that match its (currently non-credible) child nodes. When the child node eventually becomes credible, the instances collected by the parent in the child node's context during the twilight period will remain in the parent node, even though the parent can no longer make predictions in the child's context. This issue is not addressed as as it is unlikely to be of much effect.

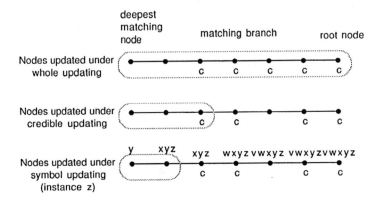

When a new instance arrives, it is added to samples in nodes on the matching branch. The SAKDC algorithm supports three methods of updating. The *whole* method (top) adds the instance to every node on the matching branch. The *credible* method (middle) adds it only to the deepest credible node and its descendants. The *symbol* method (bottom) adds the instance only to the deepest node already containing an instance of the new instance's symbol, and the node's descendents. Partial updating allows speed optimizations and improves compression. Note that the node at depth 2 is not credible. This is possible in SAKDC if both deeponly updating and windowed local adaptivity are turned on.

Figure 47: Effect of deeponly updating on the matching branch.

deeponly=symbol updating was included in the SAKDC algorithm so that Moffat's PPMC' updating[Moffat88] could be selected.

In practice, *deeponly≠whole* updating does not improve compression performance much (Section 4.17.7). This is to be expected; any context that occurs frequently will force the growth of a child node. A far more important aspect is that deeponly updating increases execution speed by reducing the number of nodes that need be updated by each arriving instance. If a depth limit is set, and the source is fairly stable, all the nodes in the tree will become credible (or contain an instance of each frequent symbol) and each arriving instance need be added only to a single node at the tip of the matching branch. Deeponly updating can be combined to great effect with shortcut pointers (Section 4.10) to avoid branch traversals altogether.

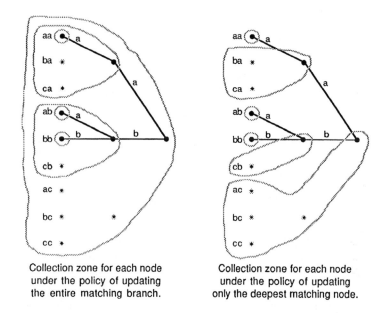

Collection zone for each node
under the policy of updating
the entire matching branch.

Collection zone for each node
under the policy of updating
only the deepest matching node.

Under *whole* updating (left), each node collects instances in contexts
matching any of its ancestor nodes. Under *credible* updating (with
a threshold of one), only the tip of the matching branch is updated.
The effect is to partition the context string space.

Figure 48: Effect of deeponly updating on node zones.

4.9 Saving Instances of Deleted Nodes

When a leaf node is moved, responsibility for making predictions in
the leaf's context falls upon its parent. This is likely to be an acceptable
degradation if the entire matching branch has received every instance.
However if deeponly updating is turned on, the parent node will have
received few of the instances that the leaf node did and will be poorly
adapted to predict in the leaf's context (**Figure 48**). By not updating
parent nodes, instances are lost when a leaf node is moved.

A solution to this problem is to record in each node the number
of instances (of each symbol) that the node, but not its parent, has
received. Then, when a node is deleted, those instances can be added
to its parents's sample. This allows node movement and deep updating
without the loss of instances. If a branch is deleted from leaf to root, the
root winds up containing all the instances in the branch.

Adding two samples together may appear to be a expensive, but turns out to have a low amortized cost. If a sparse data structure is used to represent samples, the cost of adding one sample to another is bounded by the number of symbols of positive frequency in each sample. This in turn is bounded by the number of instances in the sample. Under the addback scheme, a particular instance can be added to a parent node at most m times, after which it resides in the root node and can go no further. In practice, if deeponly updating is turned on, the rate of introduction of new instances to the tree will be about one instance per arriving instance, and the cost will be even lower.

4.10 Shortcut Pointers

Section 4.8 showed that it is unnecessary and undesirable to update the entire matching branch. Only a few (and in the limit one) node near the tip of the matching branch need be updated. Although welcome, this optimization is spoiled somewhat by the fact that the entire matching branch must be traversed to get to the node to be updated. It would be advantageous to find a method to jump among the deeper nodes without visiting the shallower nodes. Shortcut pointers provide such a mechanism.

The key observation motivating shortcut pointers is that each matching branch tightly constrains the set of possible next matching branches. More specifically, if a node whose string is $x \in S$ is in the current matching branch then the string $y \in S$ of each node in the next matching branch must share a tail of k instances with the string xa where $a \in A$ is the next arriving instance and k is the length of the shorter of the strings xa and y.

Shortcut pointers provide a mapping consisting of an *AddRight* mapping followed by zero or more *StripLeft* mappings (Section 1.12.2). A shortcut pointer associated with symbol a of node xyz could point to node xyza, node yza, node za, node a or node ϵ. The shortcut pointers are stored along with each symbol in the prediction (*AddRight*) data structure. Because the depth of the target node is not guaranteed, each node must store its own depth explicitly. The combination of the history buffer and the depth of the target node identifies the target node's string.

Figure 49 depicts a tree with shortcut pointers on the leaves only. If every node were credible and deeponly updating were turned on, the algorithm would make one transition per instance instead of the two (or in the general case m) required to move from the root to the leaf.

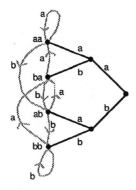

In a solid tree structure, shortcut pointers (grey) provide instant access to the tip of the next matching branch. Each node stores n shortcut pointers, one for each possible next symbol. In this tree, only shortcut pointers emerging from leaf nodes are shown.

Figure 49: Shortcut pointers in a solid tree structure.

Shortcut pointers turn the tree into a finite state machine, yielding the one-transition per instance efficiency of the DMC algorithm while retaining the flexibility and conceptual clarity of the underlying tree structure. The tree structure remains, but the shortcut pointers eliminate the costly branch traversals.

Shortcut pointers improve efficiency even when the entire branch must be updated. By providing a direct link to a node deep in the matching branch, shortcut pointers allow the matching branch to be traversed and updated from tip to root. As mentioned in Section 4.6, tip to root traversals are more efficient than root to tip traversals because tip to root traversals can use parent pointers whereas root to tip traversals require a binary search (or more generally, an n-way branch) at each level.

The example of **Figure 49** shows a solid tree. In practice the tree is more likely to be non-solid. If the "depth" of a shortcut pointer is defined to be the depth of the node that it points to, two natural constraints apply to the depth of a shortcut pointer in a non-solid tree.

- The pointer can be at most one level deeper than the source node.

- The pointer can be no deeper than the depth of the next matching branch.

Figure 50 shows a non-solid tree in which these restrictions are evident. In SAKDC, every node contains shortcut pointers (unlike the tree in **Figure 49** which contains shortcut pointers only in its leaves).

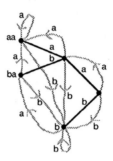

In a non-solid tree structure, shortcut pointers (grey) are constrained by the depth of the source node and cannot always point to the tip of the next matching branch. Instead, they are guaranteed only to point to *somewhere* on the next matching branch.

Figure 50: Shortcut pointers in a non-solid tree structure.

The first restriction is evident in the shortcut pointers emerging from the root node. The root node's string is the empty string, and knowledge of just one more instance allows pointers only to a depth of one, even though in the case of the a arc deeper nodes exist. The second restriction is evident in the b shortcut arc emerging from node aa. Knowledge that the next instance is b would be sufficient to point to a node "aab" of depth three. However, as such a node does not exist, the shortcut arc points to the deepest alternative which is node b.

The addition and deletion of nodes presents two problems for shortcut pointers.

If it is specified that shortcut pointers always point to the deepest node in the target branch (subject to the two constraints listed above) then a problem arises when a node is added to the tree. In **Figure 49** and **Figure 50**, all the shortcut pointers are optimally placed. If a node ab is attached to node b, the shortcut arcs b from node aa, b from ba and b from node a suddenly become sub-optimal. To update such arcs in the general case would mean modifying the b shortcut arc emerging from every node in the tree whose string ended in a.

A better solution is to optimize shortcut pointers incrementally. By removing the condition of shortcut pointer optimality from the tree's invariant, the optimization of a shortcut pointer that has been made suboptimal (by the addition of a node) can be delayed until the next time the pointer is used (**Figure 51**). The invariant for shortcut pointers becomes simply that all shortcut pointers point to *any* node on the next matching branch. A tree with all shortcut pointers pointing to the root would satisfy this invariant.

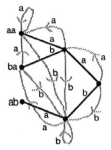 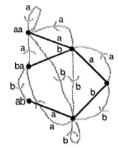

| Tree immediately after addition of the node ab. | Tree a short time afterwards. The ba-b-> shortcut has been optimized. |

In a dynamically changing tree, maintaining optimized (as deep as possible) shortcut pointers is too expensive. A better technique is to update them incrementally. Here, the addition of a new node (ab) does not immediately cause the shortcut pointers (a$\xrightarrow{b}$, aa$\xrightarrow{b}$, ba$\xrightarrow{b}$) to be updated (left). Instead, each pointer is optimized when it is next used. The diagram to the right shows the tree after the ba$\xrightarrow{b}$ pointer has been optimized.

Figure 51: Incremental shortcut pointer optimization.

To ensure that shortcut pointers are usually optimal, shortcut pointers should be tested for optimality whenever they are used, and updated if found to be suboptimal. Incremental optimizing is more efficient than eager optimizing because incremental optimization optimizes only when the information needed to optimize is readily available. Also, incremental optimization does not produce long delays (as does an $O(z)$ lumped optimization), a feature important in real-time systems. SAKDC uses incremental optimization.

A second, more serious problem arises for shortcut pointers when nodes are moved to a different part of the tree. Whereas the tree structure remains valid, shortcut pointers to a moved node become invalid and must not be used. Reconstructing an invalidated shortcut pointer is simply a matter of searching from the root. The difficult part is determining if a particular shortcut pointer is valid.

An approach to this problem is based on the fact that tree nodes that are moved need not be deallocated at the language level. In general, node movement requires only the switching of a few pointers. Thus, regardless of the volatility of the tree, each shortcut pointer is guaranteed at least to point to a valid node in the tree.[94] If deallocation never takes place, there can be no danger of a shortcut pointer ever pointing to an undefined portion of the heap.

Given that all pointers are guaranteed to point to an allocated tree node, integer incarnation number fields can be added to each node.[95] The incarnation number of each node starts at zero and is incremented whenever the node is moved. By storing incarnation numbers in shortcut pointers as well, the validity of a pointer can be determined by comparing the incarnation number of the pointer with that of the node. If the node was moved since the pointer's value was set, the incarnation numbers will differ and the pointer can be detected as invalid (**Figure 52**).

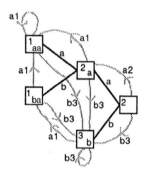

Tree before node move.

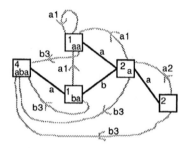

Tree after node b has been detached and reattached as node aba. Shortcut pointers labelled b3 are now invalid.

When a node is moved from one part of the tree to another, shortcut pointers that were pointing to the node become invalid, and must be detectable as such. This can be done by storing an *incarnation number* with each node and with each pointer. When a node is moved, its number is incremented. A pointer is then defined to be valid only if its number is the same as that of the node it points to. In this example, the node b changes to node **aba** and its number increases from 3 to 4. This action invalidates four pointers which remain numbered 3.

Figure 52: Incarnation numbers detect invalid shortcut pointers.

---

[94] If *local.active=true*, on occasions some allocated nodes will be in a node pool rather than in the tree (Section 4.13). As long as such nodes are not deallocated at the language level, the same argument applies (and works in practice).

[95] To the author's knowledge, the use of incarnation numbers in pointers is original.

The main disadvantages with incarnation numbers are the extra memory they use and the possibility of their overflowing. Small incarnation registers consume little memory but will overflow frequently, requiring some action. Large incarnation registers use more space but are unlikely to overflow. One encouraging fact is that under LRU, a node can be recycled at most every z instances. This means that a lower bound on the overflow time is zk instances where the incarnation register is $\lceil \log_2 k \rceil$ bits. If 16-bit registers are used, and there are 256 nodes available, it is guaranteed that overflow cannot occur until the 2^{24}'th symbol.

If an incarnation number does overflow, processing could resume after a cleanup operation during which every incarnation number in the tree is set to zero and every shortcut pointer in the tree is set to point to the root node. This would require a full tree traversal and would generate a long delay which would be unacceptable in a real-time system. Incremental solutions might exist.

4.11 Deletion of Non-Leaf Nodes

After being exposed to a stable source over a long period, the SAKDC algorithm will settle down and start to operate like a finite state machine; control will mostly flow through shortcut pointers. Under such conditions, it is tempting to organize the algorithm so as to remove the rarely used shallow non-leaf nodes for re-use as high-access leaves.

Unfortunately, abandoning the tree structure means that groups of nodes with no ancestor path to the root become hard to locate. If a node in such a group is moved, nodes that previous connected to it using shortcut pointers must find their way to the moved node's deepest ancestor. This is difficult if the target ancestor has no ancestor path to the root.

One method that might work, but which has not been explored in this thesis is to use hashing to access tree nodes. This would allow instant access to all nodes and would allow the shortcut pointers to be restored. As a further bonus, leaf nodes and non-leaf nodes could be mixed freely in an LRU system allowing a simple LRU list to be used rather than the heap system described in Section 4.7.

4.12   Credibility Thresholds

The DHPC algorithm uses a credibility threshold β to prevent samples
with fewer than β instances from being used to make predictions (Section 2.3). The SAKDC algorithm uses a credibility threshold as well (*estim.threshold.thresh*) but also provides the option (*estim.threshold.kind*)
of basing the threshold upon the *maximum* frequency in the sample
rather than the total number of instances in the sample.

The aim of a credibility test is to distinguish between samples that
are representative of the probability distribution they are modelling and
those that are not. DHPC's heuristic is based on the assumption that
samples containing at least β instances are likely to be representative.
For very large β (e.g. $\beta \geq n^2$), this is a good rule of thumb. For the
smaller values of β required in practice (e.g. $\beta \ll n$ so as to utilize the
samples of higher order nodes as early as possible), the heuristic is less
reliable.

DHPC's simple threshold scheme assumes that a fixed amount of
information (β instances) will yield approximations of the same precision
for distributions of different entropies. In fact, the higher the entropy of
the distribution, the more information is required to represent it to a
given accuracy (Section 1.10.3). For example, for $n = 256$, consider
the difference in precision of samples of ten instances from each of the
following distributions.

$$p(a) = 1/n \qquad \text{and} \qquad p(a) = \begin{cases} a = b \to 1 \\ a \neq b \to 0 \end{cases}$$

This example shows how a total-instance threshold can reject a good
approximation to a low-entropy distribution and accept a bad approximation to a high-entropy distribution. What is required is a measure
that takes the entropy of the distribution into account.

Statistical theories of confidence provide metrics for measuring how
representative a sample is. Roberts[Roberts82] has even used these in
a data compression algorithm. However, these metrics are expensive
to calculate and maintain. Here, we are concerned only with replacing
DHPC's simple, fast test with a better simple, fast test.

A threshold based upon the maximum (*max*) frequency of a sample
rather than the total (*sum*) number of instances in the sample is likely
to provide a better sample filter than *sum* thresholds because *max*
thresholds distinguish between samples based upon a crude measure of
their entropy. The lower the entropy of a distribution, the faster the

maximum frequency of a sample of increasing size is likely to increase. In the limit, the lowest average rate of increase of the maximum is $1/n$ per instance (for a high entropy distribution) and the highest is 1 per instance (for a zero entropy distribution). In contrast, the total number of instances climbs at the same rate for all entropies.

4.13   Windowed Local Adaptivity

Section 3.9 describes a simple scheme for implementing local adaptivity, in which the K'th oldest instance is removed from the tree whenever a new instance arrives. In order to reconstruct the nodes to which an instance was originally added, the history is required to be extended from K instances to $m + K$ instances. In this section we describe how this simple scheme has been integrated into SAKDC.

As mentioned in Section 3.9, if less than mK nodes are available, it is possible that growth caused by incoming instances will result in the removal of nodes from other parts of the tree and hence the destruction of instances already queued for destruction. This problem cannot be overcome by simply removing an instance from whatever nodes are present in a branch, as the end of the branch could be destroyed and then recreated during a single instance's journey through the history buffer. One solution is to reduce the rate of growth, and record the depth of each update along with each instance in the history buffer. This works for the problem posed, but does not take into account the other happenings in the more complicated algorithm (such as deeponly updating). A variable length history buffer is undesirable because the multimodal algorithm described in Chapter 5 requires a windowed locally adaptive algorithm with a fixed K.

The complete algorithm is complicated by the partial updating of branches (including deeponly updating), and the creation and destruction of nodes. Under these conditions, a more robust windowed locally adaptive scheme is required. One solution is to replace the depth of each update by a list of pointers to the updated nodes. A cyclic bounded buffer of K, $m + 1$-pointer arrays is appropriate. Each pointer contains an incarnation number which is used to prevent instances from being removed from nodes that have been moved. **Figure 53** depicts a simpler scheme in which each instance stores only a single pointer, being the the deepest node to which the instance was added.

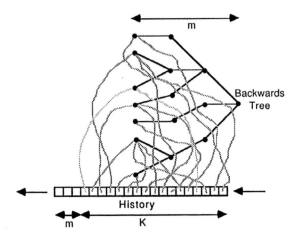

A structural *and* contextual windowed locally adaptive Markov algorithm can be implemented by storing with each instance in the history
buffer pointers to the nodes that the instance updated. When an instance reaches the end of the history buffer, it is removed from each
node that it was originally added to. If the last instance of a node is
removed, the node is deleted. This diagram depicts only the deepest
pointers.

Figure 53: Mechanism for a windowed local tree.

The only restriction on the scheme is that it cannot be combined with
decaying or addback. The scheme assumes that if a node has not been
moved since the instance (about to be removed) was added to the node,
the instance will still be present in the node's sample. Decaying and
addbacks break this invariant.

As it is possible that the number of nodes in the tree will vary, it
is necessary to maintain a pool of nodes that have been allocated but
are not currently required in the tree. This could happen if the source
is going through a low-entropy phase. It is important that this pool
be maintained so that nodes are never deallocated. This simplifies the
management of shortcut pointers (Section 4.10).

Another difficulty is the order in which nodes are recorded. If an
instance causes the creation of a whole branch, then the whole branch
may need to be destroyed when the instance is removed. This implies
that the instance must be removed from the leaf node working backwards
towards the root. Unfortunately instance addition takes place in the
opposite order. In our implementation of the SAKDC algorithm, the
$m+1$-pointers in each element of the local bounded buffer are organized
as a stack.

4.14 A Sketch of How an Instance is Processed

This section describes how the SAKDC algorithm combines the mechanisms of the previous sections into a coherent whole.

The SAKDC algorithm can be implemented as an abstract data type providing the operations *predict* and *update* (as in DHPC). SAKDC maintains a history buffer, a tree structure and a pointer to the deepest matching node, which we will call the **history node**.

The *predict* operation is performed by starting at the history node and working backwards until a credible node is found. The resultant node is the pivot node from which the estimation process is based. The search up the tree for the pivot and the possible subsequent movements up the tree (in PPM or LAZY estimation) can be performed quickly using the history node pointer and parent pointers.

The *update* operation is much messier. **Figure 54** depicts the four steps involved. Each step corresponds to a distinct movement within the tree.

Step 1: The new instance is added to nodes on the matching branch. This is done starting from the history node and moving towards the root. The *deeponly* parameter determines how far up the branch updating take place.

Step 2: The matching branch is extended according to the tree growth parameters (Section 4.3.1). Nodes that are added (if any) are updated as if they had been part of the branch all the time. No attempt is made to optimize shortcut pointers that point to the old history node. The new instance is slid into the history buffer.

Step 3: Steps 3 and 4 are concerned with establishing the (new) history node corresponding to the modified history buffer. In Step 3, the shortcut in the old history node is taken. This leads to a node somewhere on the new matching branch. If the shortcut pointer is invalid (or if shortcuts are disabled), the node will be the root node.

Step 4: The new matching branch is traversed from the position reached at the end of Step 3 to the new history node. Along the way, the shortcut pointer from the old history node is updated if

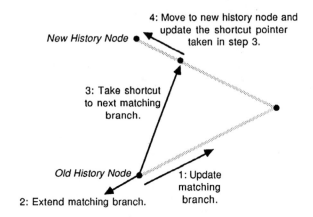

When a new instance arrives, the tree is updated in four steps. Control starts at the tip of the currently matching branch (the old history node). In Step 1, the instance is added to (some) nodes on the matching branch. In Step 2, zero or more new nodes are added to the end of the matching branch and are updated. In Step 3, the shortcut pointer in the old history node is used to jump somewhere into the new matching branch. If this is not the tip of the new matching branch, Step 4 moves to the tip and updates the shortcut pointer in the old history node. The new tip becomes the new history node.

Figure 54: Steps of an SAKDC Update.

it was suboptimal. This step yields a new history node which can be used in the next prediction and update operation.

All of the other mechanisms described in this chapter fit into this update framework.

The four movements of the four steps of an update can be described in terms of the string operators of Section 1.12.2 as 1:*StripLeft**, 2:*AddLeft*, 3:*AddRight* (and implicitly *StripLeft**) and 4:*AddLeft** where * denotes zero or more repetitions. Although the whole process may appear expensive, the process is highly efficient for a source that has any sort of stability. If *deeponly=true*, Step 1 will involve the updating of only one node. Step 2 is performed only if the depth limit has not been reached. Step 3 is expensive requiring an *AddRight* operation. Here a binary tree must be traversed to access the shortcut pointer associated with the next instance. As this traversal is necessary for coding, there is no loss here either. Finally, there are zero or more *AddLeft* operations. Again, this operation is unlikely to be necessary if the shortcut pointer is optimal.

SAKDC's complicated updating algorithm allows the use of shortcut pointers in a dynamically changing tree, and provides other access paths that speed prediction.

4.15 Controlling Implementation Complexity

At this stage, it should be clear that the SAKDC algorithm presents a minefield of potential implementation programming errors. A variety of techniques were used to control the implementation complexity.

The model was implemented as three packages:

A Prediction Package: This package exports a prediction abstraction. The abstraction allows the incrementing and decrementing of frequencies, the storage of shortcut pointers, and provides a clean interface to the coder.

A Node Package: The node package provides a tree abstraction. It is called the node package because fundamentally, it deals with nodes. A node can be detached (in which case it is not connected to anything) or it can be attached as part of a tree. The node package maintains all the information about a tree (including LRU information) and provides subprograms with information about a particular node (such as the node's depth).

A Compressor Package: The compressor package implements the actual SAKDC model. The compressor abstraction provides an operation to create a model (*create*), an operation to obtain a prediction from a model (*predict*) and an operation to give the model an instance (*update*). The compressor algorithm contains the main predict/update routines that use the prediction package and node package.

To increase reliability, assertions were sprinkled throughout each abstraction. A test package was written for the prediction package and the node package. Each test package contained a duplicate implementation of the abstraction to be tested. The duplicate implementation used simpler, but much less efficient data structures such as arrays. The test packages applied one million operations to a single instance of the abstraction with the probabilities of particular operations and data contrived to increase the probability of special cases arising frequently. For example, only seven symbols were used when testing the prediction package. Each operation was performed on the actual and simulated abstraction. If the operation involved the production of information, the information from the two abstractions was compared and an error generated if the results were different. Operations sometimes used illegal data so as to test a package's error detection behaviour. The test packages also performed consistency checks at random. No bug was ever detected in a package that had passed its test package's testing.

A test package for the compressor could not be constructed in the same manner as for the other packages as there appeared to be no way of coding the compressor's semantics any more simply than they were already coded (even at the cost of speed). No test package was used. In retrospect, a test suite should have been used.

4.16   Optimizations by Other Researchers

The author of this thesis has not been the only one involved with improving finite-context Markov algorithms. Moffat[Moffat88] has investigated ways in which PPM can be improved. Bell, Cleary and Witten summarized Moffat's improvements along with some of their own in chapter 6 of their book[Bell89].

It is interesting that every researcher who has investigated PPM has used a forwards tree whereas the author of this thesis alighted upon backwards trees. Moffat's optimizations are targetted at forward trees; however, many of them are applicable to backwards trees as well.

A comparison of past improvements with current work will be made by reviewing SAKDC's parameters. Most Markov algorithms have a depth limit (UMC and DMC being notable exceptions). PPM does not address the memory issue. PPMC' destroys the tree and rebuilds when it runs out of memory. Tree growth in nearly all other algorithms proceeds at full speed while memory is available (i.e. *grow.active=true*, *grow.threshold.thresh*=1, *grow.probext*=1.0) and stops once memory runs out (i.e. *grow.active=true*, *grow.threshold.thresh*=1, *grow.probext*=1.0). DAFC and DMC set some kind of *grow* threshold. No Markov algorithm uses any kind of LRU node management. Phoenix reconstruction is used by PPMC' with *phoenix.ashes*=2048.

Moffat used decaying in PPMC' with *decay.active=true*, *decay.threshold*=(*Sum*, 300), *decay.rounding=true* and *decay.residue=true*. Windowed local adaptivity has been used in some Huffman algorithms[Knuth85] but not in a Markov algorithm. *deeponly=credible* was invented by the author of this thesis. *deeponly=symbol* appears in PPMC'. Addbacks are original. Vine pointers (which are similar to shortcut pointers) have been used in PPM implementations. However, the suboptimality of shortcut pointers and the incarnation numbers are original. Credibility thresholds have been used by other researchers whose work has not been pursued by this author.[96] DHPC merging is really a special case of PPM merging. LAZY merging was proposed by Moffat[Moffat88] . Although none of the estimation techniques in SAKDC is original, the generalization of estimation techniques is original.

The originality of the SAKDC lies mainly in:

- its mechanisms for various kinds of adaptivity.

- its integration of generalizations of mechanisms introduced in other algorithms.

This concludes the description of SAKDC. The remainder of this chapter describes experiments that were performed to explore SAKDC's parameter space.

---

[96] See p. 19 of [Roberts82] for a reference to such work.

4.17 Experiments

This section takes up the rest of the chapter and describes the experiments performed.

4.17.1 SAKDC Implementation

The SAKDC algorithm was implemented in the Ada[USDOD83][97] programming language on a Digital VAX11/750[**Digital79**][**Digital81**] computer running VMS[**Digital78**]. Tests were run on a Vax750, a Vax780 and a Vax8530. Because of the complexity of the SAKDC algorithm, all language-based run-time checks were left on. A preprocessor called *FunnelWeb* (written by the author of this thesis) was used to generate the Ada source code. *FunnelWeb*, which is similar to the *WEB* [**Knuth83**] preprocessor, allows the programmer to weave program and documentation together. This improves code readability and reliability.

SAKDC (like DHPC) is best described as a process that reads a stream of instances and writes a stream of predictions. SAKDC implements the model unit of **Figure 9**. Section 2.4.1 explains how such a model can be used in a compression program. In our implementation, the SAKDC algorithm drives an arithmetic coder. However, the coder was so slow (partly because of the double precision floating point — see Section 4.4) that it was turned off for most of the experiments. Compression was calculated from the sum of the entropies of the instances processed. In all cases when the coder was turned on, compression was within a few bytes[98] of the theoretically calculated figure.

The SAKDC implementation used in these experiments sets n (the number of symbols in the set of symbols) to 256 and divides the input file into a stream of bytes each of which is treated as a separate instance.

4.17.2 Description of the Test Data

Researchers in data compression at the University of Calgary have prepared and released a corpus of "standard" files to be used in data compression research.[99] This is an enlightened move, because it allows

---

[97] Ada is a registered trade mark of the US-Government-Ada Joint Program Office.
[98] Within six bytes; the coder register was six bytes wide.
[99] The author of this thesis received his copy of the corpus from Ian Witten and Tim Bell on 4 February 1988. Corrections to the *pic* file were received and made on 14 March 1988.

data compression researchers in many countries to compare the performance of their data compression algorithms on a standard set of data. We will call the corpus of files from Calgary the "Calgary corpus".

The corpus of files assembled to test SAKDC consisted of the Calgary corpus with some additional files added to assist in the investigation of adaptivity. These extra files were necessary because many of the files in the Calgary corpus are fairly homogenous.

Table 8 lists the files that were used in the experiments to be described. To avoid ambiguity, each file has been given a name and a number. The names given to files in the Calgary corpus are the same as those used by the originators of the corpus. Files numbered in the range $Z[1, 18]$ are members of the Calgary corpus. The remaining files ($Z[101, 104]$) were added by the author of this thesis.[100] Checksums were calculated according to the following formula, with each file being treated as a sequence of eight-bit bytes in the range $Z[0, 255]$.

$$checksum(s) = \begin{cases} |s| = 0 \to 0 \\ |s| > 0 \to \left(3 \times checksum(s_{1...|s|-1}) + s_{|s|} + 1\right) \bmod 2^{16} \end{cases}$$

Further information about files in the Calgary corpus can be found in [Bell89]. The files added to the corpus by the author of this thesis are all files that change their characteristics in interesting ways. The *multi* file consists of the output of three simple, pure Markov sources interleaved at 10000-instance intervals. The *concat* file was constructed by concatenating four files together and then concatenating the result to itself. The four files consisted of PostScript code, numeric data in text form, a hex dump and a program written in an editor language. The file *inter* consists of twenty 10000-instance chunks from four corpus files and the ASCII numeric file. The *objpap* file is simply the *obj1* file followed by the *paper1* file. A more detailed description of the files *multi*, *concat* and *inter* can be found in Section 5.7.

4.17.3 Presentation of Experiments

A wide range of experiments were performed to explore SAKDC's parameter space. A quick review of the literature revealed that data compression experiments are usually described in a conversational style. While highly readable, this style suffers from the following maladies.

---

[100] If any sort of standard arises from all this, the author proposes that all files be given a unique name and number, that Calgary be allocated the range $Z[0, 99]$, and that the author be allocated the range $Z[100, 199]$.

Num	Name	Length	Chksum	Text	Description
1	*bib*	111261	57927	Yes	A bibliography
2	*book1*	768771	49516	Yes	A fiction book
3	*book2*	610856	4041	Yes	A non-fiction book
4	*geo*	102400	44972	No	Geophysical data
5	*news*	377109	5454	Yes	A News batch file
6	*obj1*	21504	38609	No	A Vax object file
7	*obj2*	246814	57595	No	A Mac object file
8	*paper1*	53161	2278	Yes	A paper
9	*paper2*	82199	10028	Yes	A paper
10	*paper3*	46526	29255	Yes	A paper
11	*paper4*	13286	58710	Yes	A paper
12	*paper5*	11954	27661	Yes	A paper
13	*paper6*	38105	43918	Yes	A paper
14	*pic*	513216	7886	No	A CCITT test image
15	*progc*	39611	27350	Yes	A C program
16	*progl*	71646	33011	Yes	A Lisp program
17	*progp*	49379	47806	Yes	A Pascal program
18	*trans*	93695	48136	Yes	A terminal session
101	*multi*	90000	18527	No	Multimodal (see text)
102	*concat*	185286	58496	Yes	Many files (see text)
103	*inter*	200000	2911	No	Many files (see text)
104	*objpap*	74665	50617	No	*obj1+paper1*

This table lists the data files used in experiments in this thesis. Files numbered $Z[1, 18]$ form a corpus of files prepared by researchers at the University of Calgary. The other files were prepared by the author of this thesis and contain data generated by a moving source.

Table 8: The corpus of data files used to test SAKDC.

- Where there is more than one algorithm, it is not made clear *which* algorithm is being described.

- The values of some algorithm parameters used in experiments are omitted.

- The nature of the data, and in particular the length, is not adequately specified.

- Discussion of the experiments drifts from experiment to experiment with no clear indication of whether the mechanisms tested in one experiment are being employed in the next.

- The discussion is not amenable to "random access". That is, a reader cannot obtain all the details and the conclusion of a particular experiment without reading the entire text.

To avoid these difficulties, we revert to the classic scientific experimental writing up style of *aim, method, results, conclusions*. To the extent that SAKDC has a large multidimensional parameter space, it can be considered to be a natural phenomenon worthy of experimental observation.

Unless otherwise specified, the vertical axis of all the graphs in the following experiments measures compression in the form "proportion remaining" (Prop.Rem.). The measure "proportion remaining" was chosen over "bits/instance" because of its direct practical applicability in a world of eight-bit bytes. Most people (including the author of this thesis) are skilled at quantification in base ten but not in base eight. Despite this rejection, bits/instance remains the most objective measure of compression.[101]

In each experiment, a table of SAKDC **base parameters** will be given. The effect of varying one or more of the parameters (the "independent variable") under those conditions will then be described. Parameters listed as having a value of "?" in the base parameter table are the independent variables. The *shortcuts* parameter is not listed as it has no effect on compression. Unless otherwise stated, *shortcuts=true*. As stated earlier, $n = 256$ in all runs.

Of the experiments to be described the most important are:

Experiment 2 which evaluates different estimation formulae.

Experiment 5 which evaluates the effect of depth.

Experiment 6 which evaluates the effect of memory.

Experiment 12 which evaluates the effect of structural adaptivity.

---

[101] Except of course for nats/instance!

4.17.4 Experiment 1: Initial Benchmarks (DHPC, PPMA, PPMC')

Aim: To determine the performance of some well-known algorithms.

Method: Three methods were chosen for testing: DHPC, PPMA and PPMC'. PPMA is a variation of PPM; the "A" describes the estimation technique. PPMC' is a refinement of PPM by Moffat[Moffat88]. PPMC' is also described in [Bell89]. The parameters chosen for each method are as close to those specified by their inventors as could be determined from their papers. The parameters for DHPC are given in **Table 9**. The parameters for PPMA are given in **Table 10**. The parameters for PPMC' are given in **Table 11**. Each method was run over the entire corpus of files.

Parameter	Value	Parameter	Value
Maxdepth	4	*Maxnodes*	20000
Grow	Yes, Sum of 2, Pext=1.0	*Move*	No
Lruparent	Same	*Phoenix*	No
Local	No	*Decay*	No
Deeponly	Whole	*Addback*	No
Estim	DHPC, Lin, $\lambda = 1$	*Estim.threshold*	Sum of 3

Table 9: Experiment 1: Benchmark parameters for DHPC.

Parameter	Value	Parameter	Value
Maxdepth	4	*Maxnodes*	20000
Grow	Yes, Sum of 1, Pext=1.0	*Move*	No
Lruparent	Same	*Phoenix*	No
Local	No	*Decay*	No
Deeponly	Whole	*Addback*	No
Estim	PPM, NonLin, $\lambda = 1$	*Estim.threshold*	Sum of 1

Table 10: Experiment 1: Benchmark parameters for PPMA.

Parameter	Value	Parameter	Value
Maxdepth	3	*Maxnodes*	20000
Grow	Yes, Sum of 1, Pext=1.0	*Move*	No
Lruparent	Same	*Phoenix*	Yes, Ashes=2048
Local	No	*Decay*	Yes, Sum of 300, 0.5, Rou, Res
Deeponly	Symbol	*Addback*	No
Estim	PPM, NonLinMof, $\lambda = 1$	*Estim.threshold*	Sum of 1

Table 11: Experiment 1: Benchmark parameters for PPMC'.

Results: The results are listed in **Table 12**. DHPC is inferior to PPMA except for the files *geo*, *obj1*, *obj2*. PPMA is inferior to PPMC' except for the files *book1* and *pic*.

Num	Name	DHPC	PPMA	PPMC'
1	*bib*	0.315	0.268	0.264
2	*book1*	0.337	0.304	0.315
3	*book2*	0.312	0.289	0.288
4	*geo*	0.732	0.761	0.671
5	*news*	0.403	0.374	0.350
6	*obj1*	0.586	0.626	0.497
7	*obj2*	0.435	0.453	0.348
8	*paper1*	0.389	0.328	0.311
9	*paper2*	0.376	0.314	0.307
10	*paper3*	0.432	0.354	0.338
11	*paper4*	0.487	0.406	0.366
12	*paper5*	0.498	0.427	0.381
13	*paper6*	0.402	0.343	0.318
14	*pic*	0.123	0.112	0.161
15	*progc*	0.391	0.345	0.315
16	*progl*	0.286	0.247	0.239
17	*progp*	0.275	0.247	0.235
18	*trans*	0.257	0.225	0.223
—	average	0.391	0.357	0.329

This table lists the compression performance of the algorithms DHPC, PPM and PPMC'. Compression is expressed as the proportion remaining. Although, the three algorithms yield similar compression, PPMC' emerges as the clear winner.

Table 12: Experiment 1: Benchmarks for DHPC, PPMA and PPMC'.

DHPC's better performance on three files (files that could be expected to contain changing data) could be because of its slower tree growth. In the *geo* run (of 102400 instances), PPMA stopped growing its tree after 10755 instances whereas DHPC stopped after 20986 instances. It could also be the differences in estimation scheme.

DHPC ran at about 120 instances per CPU second, PPMA at about 80 per CPU second and PPMC' at about 100 instances per CPU second. These figures are rough as no attempt was made to optimize the program.

Conclusions: The methods DHPC, PPM (PPMA) and PPMC' yield results that are within a few percent (absolute) of each other. However, it is safe to say that PPMC' out-performs PPM (by about 3% absolute) and PPM out-performs DHPC (by about 3% absolute).

4.17.5 Experiment 2: Estimation (*estim*)

Aim: To evaluate the performance of different estimation techniques.

Method and Results: Although it would be desirable to test each SAKDC parameter independently of the others, some parameters are so closely related that they must be tested together. This experiment tests the parameter group *estim*. The *estim.threshold* parameter is held constant at 1 so as to expose the estimation techniques to small samples; because all estimation techniques must converge as the sample size approaches infinity (Section 1.10.3), differences between the estimation techniques are likely to be most evident for small sample sizes.

The compressor runs in Experiment 1 took six hours of CPU time on a VAX750 for each run through the corpus. This is about two hours per megabyte. To speed things up, three small files were selected for further experimentation. These were *obj1*, which contained object code, *paper1*, which contained English text, and *progc*, which contained a C program. This amounted to 112K which could be run in a total of about 15 minutes of CPU time.

Parameter	Value	Parameter	Value
Maxdepth	4	*Maxnodes*	20000
Grow	Yes, Sum of 1, Pext=1.0	*Move*	No
Lruparent	Same	*Phoenix*	No
Local	No	*Decay*	No
Deeponly	Whole	*Addback*	No
Estim	?, ?, λ =?	*Estim.threshold*	Sum of 1

Table 13: Experiment 2: Estimation base parameters.

The experiment commenced with a run of eight different estimation techniques for the file *paper1* over a wide range of λ. The base parameters used for the runs in this experiment are listed in **Table 13**. The results of the first "wide angle" run are shown in **Figure 55**.

In this and further graphs in this experiment, circles are used for linear estimation and squares for non-linear estimation. A dotted line indicates that the Moffat variant was used. Thick lines are used for PPM merging, medium lines for LAZY merging, and thin lines for DHPC merging. The horizontal axis measures λ and the vertical axis measures compression.

As might be expected, **Figure 55** shows that the best values for λ are fairly small (< 10). **Figure 56** shows the more interesting range $R[0, 5]$.

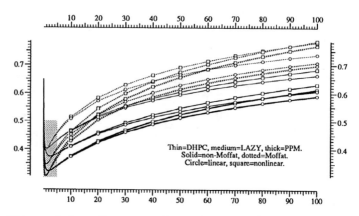

This graph shows the compression performance (y-axis) of the twelve different combinations of merging and estimation techniques over a wide range of λ (x-axis) for the file *paper1*. This graph shows that small values of λ are best. The best compression is obtained in the shaded region, which is shown in greater detail in **Figure 56**.

Figure 55: Experiment 2: Prop.Rem. vs λ for *paper1* (wide angle).

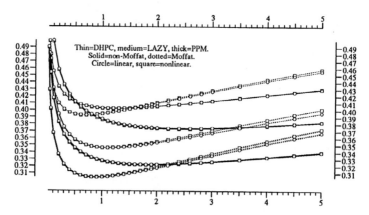

This graph corresponds to the shaded region of **Figure 55** and gives a better idea of the best estimation techniques. Linear (circles) and non-linear (squares) estimation techniques perform nearly identically. PPM merging is better than LAZY merging which is better than DHPC merging. In each case, Moffat's modification improves compression considerably.

Figure 56: Experiment 2: Prop.Rem. vs λ for *paper1* (close up).

It is immediately clear from **Figure 56** that the linear (circles) and non-linear (squares) techniques perform identically over this range. Indeed they are so close that only six or so lines appear on the graph even though twelve were plotted. The squares of the lines of the nonlinear methods (plotted underneath the linear methods) can be seen at the very right of the graph where the lines diverge slightly. **Figure 55** shows that the linear and non-linear line pairs diverge by up to 7% at $\lambda = 100$. Because linear and non-linear methods are so similar in effect, only linear methods will be used in further runs. When designing a production data compressor, the choice of linear or non-linear should be determined by ease of implementation.

Another prominent aspect of **Figure 56** is the superiority of PPM merging over DHPC merging and LAZY merging.

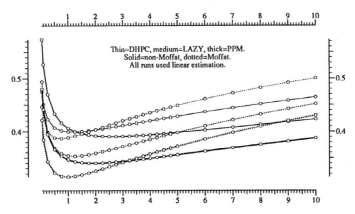

This graph is very similar to **Figure 56** and suggests the same conclusions.

Figure 57: Experiment 2: Prop.Rem. vs λ for *progc*.

To ensure that the results for *paper1* were not file specific, the same run was performed on the files *progc* and *obj1*. The λ range was extended to $R[0,10]$. The results for *progc* (**Figure 57**) are much the same as for *paper1*. For *obj1* (**Figure 58**) the curves are quite different. The object file has a higher entropy than the other files and the curves are situated higher. The graph has split into Moffat and non-Moffat line groups instead of PPM, LAZY and DHPC groups. Non-Moffat DHPC merging performs better here than non-Moffat PPM merging. All the curves have slumped right (relative to **Figure 56**), and the optimum λ has followed. However, because of the flatness of the slump, low values of λ would still yield near optimal performance.

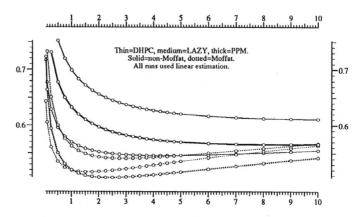

This graph shows the performance of various merging and estimation techniques for a range of λ on an object file. In this graph, the line groups split into Moffat and non-Moffat rather than DHPC, LAZY and PPM groups.

Figure 58: Experiment 2: Prop.Rem. vs λ for *obj1*.

The next two runs tested the estimation techniques for few and many instances. To obtain samples with few instances, the model was run on *paper1* with the *move* parameters set the same as the *grow* parameters and *local.active=true* and *local.period*=500, meaning that the model would forget all but the last 500 instances (**Figure 59**). Apart from raising the curves by about 15%, the result is little different to **Figure 56**.

To obtain samples with many instances, the compressor was run with *maxdepth*=1 on *paper1* (**Figure 60**) and *obj1* (**Figure 61**). For *paper1*, the effect of lowering the depth (and hence increasing the sample size) is to flatten the curves (observe the y axis labellings). Strangely, the optimal λ for the non-Moffat formulas has *increased* while that of the Moffat formulas has slightly decreased. For the object file, the curves have moved up and retained their shape, although the PPM, LAZY and DHPC pairs have moved closer.

During the above runs, statistics were accumulated, after each prediction was made, of the sample of the deepest node used to make the prediction. **Table 14** shows the average number of instances (*AvNumIns*), the average maximum frequency (*AvMaxFrq*) and the average number of symbols (*AvNumSym*) in the predictions. The table confirms that turning on windowed local adaptivity and lowering the depth were

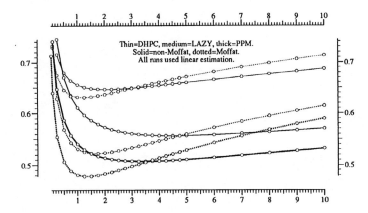

This graph shows the performance of various merging and estimation techniques where the number of instances in samples has been held low by setting *local.period*=500. These results are similar to earlier results.

Figure 59: Experiment 2: Prop.Rem. vs λ for *paper1* with *local.period*=500.

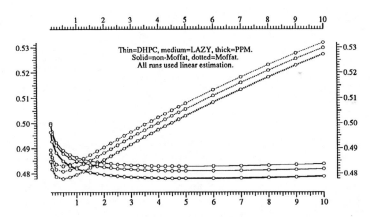

This graph shows the performance of various merging and estimation techniques where the number of instances in samples has been made artificially high by setting the maximum depth to one. Again, PPM merging and Moffat estimation yield the best performance, although the curves have become much closer.

Figure 60: Experiment 2: Prop.Rem. vs λ for *paper1* with *maxdepth*=1.

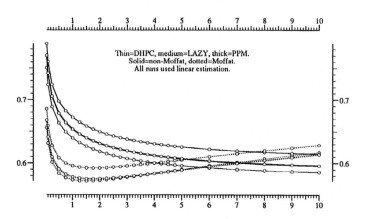

This graph shows the performance of various merging and estimation
techniques where the number of instances in samples has been made
artificially high by setting the maximum depth to one. For the object
file, DHPC yields the best compression.

Figure 61: Experiment 2: Prop.Rem. vs λ for *obj1* with *maxdepth*=1.

good ways of manipulating the instance density. The 4.15 entry was unexpected but can be explained by the decreasing depth of predictions as
the supply of nodes decreases.

Run	Figure	*AvNumIns*	*AvMaxFrq*	*AvNumSym*
Few instances	**Figure 59**	19	4.5	4.15
Ordinary	**Figure 56**	32	14.1	3.75
Many instances	**Figure 60**	1339	289	30

This table lists average statistics of the samples of the deepest nodes
used to make predictions in each run. *AvNumIns* is the average
number of instances in the samples. *AvMaxFrq* is the average
maximum frequency in the samples. *AvNumSym* is the average
number of symbols with positive frequency in the samples. These
results show that the use of windowed local adaptivity and low depths
were effective at manipulating instance density in these runs.

Table 14: Experiment 2: Symbol and instance densities.

Conclusions: The optimum λ for all the estimation techniques nearly
always lies in the range 0 to 5. The linear and non-linear forms yield
almost identical performance for $0 < \lambda < 10$ but diverge for $\lambda > 10$
with linear yielding better performance. The Moffat forms of estimation
with $\lambda \approx 1$ perform significantly better than the non-Moffat forms.

PPM merging performs 5% to 10% better than DHPC merging unless the entropy of the source is high, in which case DHPC merging does a few percent better than PPM merging. The performance of the LAZY merging technique is roughly between that of DHPC and PPM merging. The best estimation technique is PPM merging with Moffat estimation with $\lambda = 1$ (linear or non-linear).

4.17.6 Experiment 3: Credibility (*estim.threshold*)

Aim: To determine the effect on compression of credibility thresholds.

Method: In this experiment the compressor was run over the files *paper1* and *obj1* for a range of values of *estim.threshold*. All six estimation techniques were used (linear estimation is used exclusively in the remaining experiments). The base parameters are listed in **Table 15**.

Parameter	Value	Parameter	Value
Maxdepth	4	*Maxnodes*	20000
Grow	Yes, Sum of 1, Pext=1.0	*Move*	No
Lruparent	Same	*Phoenix*	No
Local	No	*Decay*	No
Deeponly	Whole	*Addback*	No
Estim	?, LinMof, $\lambda = 1$	*Estim.threshold*	? of ?

Table 15: Experiment 3: Credibility base parameters.

Results: The results for *paper1* are shown in **Figure 62**. The graph shows clearly that use of any sort of credibility threshold is detrimental to compression unless predictions are based solely on the deepest node (DHPC merging).

Figure 63 leads to the same disappointing conclusion. It appears that it is better to blend in *all* the available information rather than to cut it off arbitrarily.

> "In theory, theory is the same as practice, but in practice it isn't." Fortune Cookie

Conclusions: Credibility thresholds are detrimental to compression and should not be used (i.e. *estim.threshold.thresh* should be set to 1). An exception is when predictions are to be based entirely on the deepest matching node (DHPC merging) in which case, a *max* threshold of 3 will improve compression by a few percent absolute.

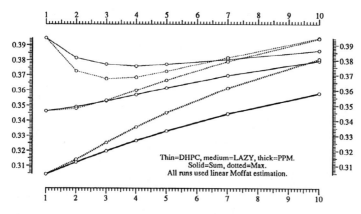

This graph shows the compression performance (y-axis (proportion remaining)) of three different merging techniques (using linear Moffat estimation with $\lambda = 1$) over a range of credibility thresholds (x-axis) for *paper1*. Dotted lines are used for *max* thresholds and solid lines for *sum* thresholds. Thresholding improves compression only for DHPC and then only a little. Generally estimation thresholding appears to be detrimental to compression.

Figure 62: Experiment 3: Prop.Rem.
vs *estim.threshold.thresh* for *paper1*.

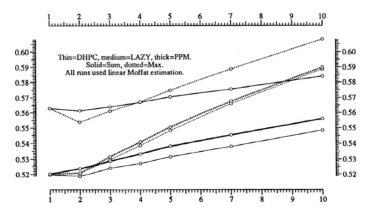

This graph shows the compression performance (y-axis (proportion remaining)) of three different merging techniques (using linear Moffat estimation with $\lambda = 1$) over a range of credibility thresholds (x-axis) for *obj1*. Dotted lines are used for *max* thresholds and solid lines for *sum* thresholds. Thresholding improves compression only for DHPC and then only a little. Generally estimation thresholding appears to be detrimental to compression.

Figure 63: Experiment 3: Prop.Rem.
vs *estim.threshold.thresh* for *obj1*.

4.17.7 Experiment 4: Deep Updating (*deeponly*)

Aim: To evaluate the performance of different updating methods (*deeponly*).

Method: The compressor was run over the entire corpus for each value of the *deeponly* parameter. The base parameter values are listed in Table 16.

Parameter	Value	Parameter	Value
Maxdepth	4	*Maxnodes*	20000
Grow	Yes, Sum of 1, Pext=1.0	*Move*	No
Lruparent	Same	*Phoenix*	No
Local	No	*Decay*	No
Deeponly	?	*Addback*	No
Estim	PPM, LinMof, $\lambda = 1$	*Estim.threshold*	Sum of 1

Table 16: Experiment 4: Deep updating base parameters.

Results: The results of the runs are shown in **Table 17**.

The table shows that *symbol* out-performed *credible* and *whole* updating by about 2% absolute. *whole* out-performed *credible* for every file except the object files *obj1* and *obj2* for which *whole* performed about 1% absolute worse. Moffat[Moffat88] found that *symbol* updating yields a 5% relative improvement over *whole*. Here it is 3.7% relative.

Although the compressor program was not much optimized, it is instructive to review the execution speeds. The speed of these runs (expressed in bytes per second on a Vax8530 and averaged in the same manner as the compression results of **Table 17**) were *whole*:404, *credible*:506 and *symbol*:487. It seems that any sort of modification that prevents shallow nodes from being updated will increase execution speed by about 25%.

Conclusions: Assuming PPM estimation, the best updating technique (value for the *deeponly* parameter) is *symbol*. That is, it is best to update only the matching descendents of the deepest matching node whose sample already contains at least one instance identical to that about to be added. *Symbol* yields a compression improvement of about 1% absolute (4% relative) over *whole* while speeding the compressor up by about 25%. *Credible* (adding the instance only to the deepest matching node (*estim.threshold*=1)) yields the same speed increase but loses about 1% absolute in compression.

Num	File	Whole	Credible	Symbol
1	*bib*	0.253	0.264	0.244
2	*book1*	0.299	0.307	0.296
3	*book2*	0.281	0.281	0.270
4	*geo*	0.605	0.626	0.586
5	*news*	0.351	0.354	0.337
6	*obj1*	0.520	0.516	0.481
7	*obj2*	0.424	0.408	0.403
8	*paper1*	0.304	0.326	0.296
9	*paper2*	0.299	0.323	0.294
10	*paper3*	0.330	0.358	0.325
11	*paper4*	0.372	0.406	0.362
12	*paper5*	0.386	0.420	0.375
13	*paper6*	0.316	0.341	0.305
14	*pic*	0.103	0.107	0.103
15	*progc*	0.314	0.334	0.302
16	*progl*	0.234	0.235	0.221
17	*progp*	0.230	0.245	0.221
18	*trans*	0.211	0.215	0.200
—	average	0.324	0.337	0.312

This table lists the performance of *whole*, *credible* and *symbol* updating on the files in the Calgary corpus. The compression performance of the different methods is similar, but *symbol* emerges as the clear winner by a few percent.

Table 17: Experiment 4: Deep updating results.

4.17.8 Experiment 5: Depth of Tree (*maxdepth*)

Aim: To determine the effect of tree depth on compression.

Method: The compressor was run for *maxdepth* values $\mathbf{Z}[1, 10]$ over five different files: *geo*, *obj1*, *paper1*, *progc* and *bib*. The base parameters for the runs are shown in **Table 18**.

Parameter	Value	Parameter	Value
Maxdepth	?	*Maxnodes*	20000
Grow	Yes, Sum of 1, Pext=1.0	*Move*	No
Lruparent	Same	*Phoenix*	No
Local	No	*Decay*	No
Deeponly	Symbol	*Addback*	No
Estim	PPM, LinMof, $\lambda = 1$	*Estim.threshold*	Sum of 1

Table 18: Experiment 5: Depth base parameters.

Results: The results are shown in **Figure 64**. The graph shows that increasing the tree depth helps, but only to an extent. Increases up to order 4 have a huge effect after which increasing the depth appears to be mildly detrimental to compression. The graph also shows the different complexities of the different files. The *geo* file seems to contain only first-order correlations whereas *bib* contains at least fourth-order correlations.

One explanation for the lack of effect of extra depth after depth 4 is that when memory runs out, deeper nodes are more profitable employed at shallower levels. **Table 19** lists for each file and depth, the rate of growth in nodes/instance during the period that the tree was growing. The table shows that each extra level of depth increases memory consumption considerably.

Table 20 lists, for each run, the number of instances before memory ran out. To an extent, the figures here correlate with the decrease in performance at depth 4. However, there is also evidence to suppose that memory is not the governing factor here. For *geo*, the runs at depth 1 and 2 did not run out of memory and yet there is little difference in their performance. Similarly, for *obj1* for depths 2 and 3 and *paper1* and *obj1* for depths 3 and 4. From these results it appears that if memory is available, extra depth will yield minor improvements but that there is not much to gain after depth 4. Because of the huge difference in memory consumption between depths 3 and 4, depth 3 is recommended.

Conclusions: Starting from depth 0, compression improves quickly (typically yielding an extra 10% absolute per extra level) up to and including depth 3. Above depth 3, compression can improve by up to 3% but is more likely to degrade. A depth of 3 is recommended.

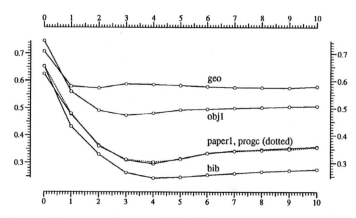

Along with memory, the depth of the Markov tree is a most important determinant of compression performance. Here the compression performance (proportion remaining) is plotted against tree depth for five files. Although for small depths compression improves dramatically with increasing depth, there appears to be no advantage in increasing the depth of the tree beyond depth 4. A depth of 3 is preferred because of its decreased memory consumption.

Figure 64: Experiment 5: Prop.Rem. vs *maxdepth* for files.

Depth	*geo*	*obj1*	*paper1*	*progc*	*bib*
1	0.0025	0.0120	0.0018	0.0023	0.0007
2	0.138	0.240	0.031	0.046	0.015
3	0.78	0.66	0.015	0.19	0.87
4	1.85	1.30	0.40	0.48	0.38
5	2.86	1.71	1.10	1.36	1.07

Table 19: Experiment 5: Rate of node growth for different depths.

Depth	geo	obj1	paper1	progc	bib
1	102400	21504	53161	39611	111261
2	102400	21504	53161	39611	111261
3	25596	21504	53161	39611	111261
4	10755	15344	49869	39611	52425
5	6970	11728	18057	14750	18668

Table 20: Experiment 5: Memory run out times for different depths.

4.17.9 Experiment 6: Memory Size (*maxnodes*)

Aim: To determine the effect on compression of memory size (*maxnodes*).

Method: In this experiment memory is measured in tree nodes rather than in bytes. This makes the measure portable as well as easy to constrain. The compressor was run over a range of tree sizes for the files *paper1*, *obj2* and *trans*. The runs were performed for depths of 3 and 4 so as to confirm the results of Experiment 5.

Parameter	Value	Parameter	Value
Maxdepth	?	*Maxnodes*	?
Grow	Yes, Sum of 1, Pext=1.0	*Move*	No
Lruparent	Same	*Phoenix*	No
Local	No	*Decay*	No
Deeponly	Symbol	*Addback*	No
Estim	PPM, LinMof, $\lambda = 1$	*Estim.threshold*	Sum of 1

Table 21: Experiment 6: Memory base parameters.

Results: The results are shown in **Figure 65**. Triangles mark the depth 3 runs and squares mark the depth 4 runs.

Starting from just one node, compression improves dramatically as nodes are added. For the file *trans*, a tree of 500 nodes yields 20% absolute more compression than a tree with just one node. Adding another 500 removes another 5% absolute.

Zooming out, **Figure 66** shows a wider view. The shaded region contains Figure 65. As might be expected, the improvement in compression diminishes as more nodes are added. Here the curve trails off at about 10000 nodes.

It is interesting that a depth of 4 yields slightly better compression (2% absolute) if there are more than (approximately) 12000 nodes available. This indicates that depth 4 nodes are a luxury that can be afforded only when nodes are plentiful. The points at the far right of the graph plot the position of the curves at 60000 nodes. For the *paper1* and *trans* files, compression does not improve with extra memory beyond about 20000 nodes. For *obj2*, there is a 5% absolute improvement. This is probably because the *obj2* file is longer than the other files. The graph shows that a depth 4 tree does not outperform a depth 3 tree even if memory is tripled. It is likely that a depth 2 tree would outperform a tree of depth 3 under even more stringent memory constraints.

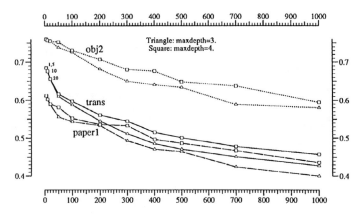

Along with tree depth, memory is a most important determinant of compression performance. Here the compression performance (proportion remaining) is plotted against *maxnodes* (the maximum number of nodes allowed in the tree) for three files. Depths of 3 and 4 are used to confirm the results of Experiment 5, and they do, with depth 3 performing as well or better than depth 4. When memory is small, compression improves dramatically with increasing memory. In this case, increasing memory from 1 to 1000 nodes has improved compression by 30% (relative).

Figure 65: Experiment 6: Prop.Rem. vs *maxnodes* for files (close up).

Conclusions: Starting with one node, huge gains in compression can be made by increasing the number of nodes in the tree. If one node yields 0.7 (proportion remaining) then 500 nodes will yield 0.5. At 5000 nodes, compression tapers off and by 10000 nodes there is little to be gained by increasing memory. These results confirm the optimal depth as 3. For less than 500 nodes, decreasing the depth can improve compression. For over 20000 nodes, compression can improve for a depth of 4 but is unlikely to improve for higher depths. These results are based on ordinary files of ordinary length.

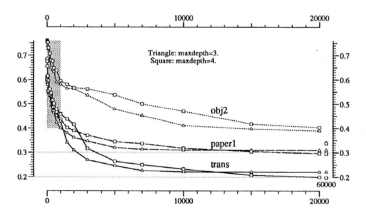

This graph gives a wider view of the data given in **Figure 65** (enclosed in the grey region). This graph shows that while great improvements in compression can be obtained by adding memory when memory is scarce, there is little advantage in increasing memory above 10000 nodes.

Figure 66: Experiment 6: Prop.Rem.

vs *maxnodes* for files (wide angle).

4.17.10 Experiment 7: Windowed Local Adaptivity (*local*)

Aim: To determine the locality of typical data and consequently the best windowed local adaptivity settings (*local*).

Method: For this experiment the SAKDC algorithm's windowed locally adaptive mechanism was used (*local*). Unlike the decay mechanism, the windowed locally adaptive mechanism places a fixed bound on the maximum age of instances in the tree.

The compressor was run with the parameters listed in **Table 22** for a variety of values of *local.period*. The compressor was given 50000 nodes so as to lower the chance of it having to recycle nodes prematurely (for a depth of 3 and 50000 nodes, premature recycling could only happen for *local.period* $> \approx$ 16000). Six files were tested: *obj1*, *obj2*, *paper1*, *progc*, *book1* [1 ... 200000], *multi* and *inter*. By "*book1* [1 ... 200000]" is meant the first 200000 instances of the file *book1*; performing the runs on the whole book would have taken too long.

Parameter	Value	Parameter	Value
Maxdepth	3	*Maxnodes*	50000
Grow	Yes, Sum of 1, Pext=1.0	*Move*	Yes, Sum of 1, Pext=1.0
Lruparent	Same	*Phoenix*	No
Local	Yes, Period=?	*Decay*	No
Deeponly	Symbol	*Addback*	No
Estim	PPM, LinMof, $\lambda = 1$	*Estim.threshold*	Sum of 1

Table 22: Experiment 7: Windowed adaptivity base parameters.

Results: The results of the run are listed in **Figure 67**. The horizontal axis plots *local.period* and the vertical axis plots compression (proportion remaining). The fuzzy horizontal lines are included solely to provide a visual reference of the horizonal at interesting points on the graph. The point markers (squares, circles, triangles) have no significance other than to distinguish the different lines. Dotted lines were used for non-homogenous files.

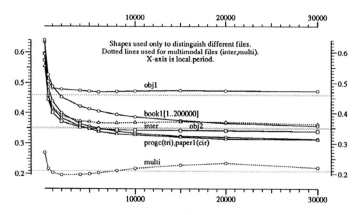

Section 4.13 described how windowed local adaptivity could be implemented by removing instances from the tree when they are K instances
old. Here compression performance (proportion remained) is plotted
against the length of the window (*local.period*) for a variety of files.
With the exception of the multimodal files (dotted), compression is
improved by increasing the size of the window. This indicates a poor
locality in ordinary data. Because compression is not much worse for
smaller windows, small windows could be used in practice to simplify
implementations.

Figure 67: Experiment 7: Prop.Rem. vs *local.period* for some files.

Of the homogeneous files (*obj1*, *obj2*, *book1* [1 . . . 200000], *progc* and
paper1), only the object files (*obj1* and *obj2*) have a low optimum
locality. The rest all appear to have optimum localities at infinity.
This result contrasts strongly with the generally held view that data
is highly localized (e.g. [Bentley86], [Abrahamson89]). The difference
might be caused by the level at which the data is parsed (Bentley —
words, Abrahamson — bytes), by the order of the model (Bentley —
0, Abrahamson — 1) or by the differing w (Section 3.3) which in this
experiment is flat but in the other research was strongly biased towards
the most recent instances.

Of great interest is the fact that for most of the files, only 5000 to
10000 instances were required to give good compression. Arithmetic
coding becomes clumsy if frequency counts can grow large (e.g. > 16383
([Moffat88])). Typically decaying is used to keep these counts down.
However, as only about 10000 instances are needed in a compressor to
produce reasonable compression, windowed local adaptivity could also
be used to place an upper bound on the counts.

Of all the files tested, *book1* appears to be the most homogeneous with a 4% gap between 10000 and 30000 nodes.

The results of the multimodal files *multi* and *inter* are also surprising. As expected, the optimum period for *multi* is about 3000 instances. Any more, and the advantage gained by keeping extra instances of the current source are lost at the source boundary. What is surprising is the downwards dip in the curve at 30000. This figure can be compared to the period of the repetitions of the three sources in *multi* which is also 30000. The *inter* curve shows the same effect although it is less dramatic.

Conclusions: For byte-level high-order models with flat w, the optimum sized window is one of infinite length. However, only the most recent 5000 to 10000 instances are needed to achieve near optimal compression. This fact could be used to keep frequency counts down, simplifying the design of practical arithmetic coders. The result also means that there is little to lose (say 5%) by making a compressor locally adaptive. However, if compression is the absolute priority then asymptotically adaptive models will be best for highly stable sources.

4.17.11　Experiment 8: Decaying (*decay*)

Aim: To determine the effect on compression of sample decaying.

Method: The compressor was run with the parameters listed in Table 23 for a variety of values of *decay.threshold*. The other decay parameters (*decay.factor*, *decay.rounding* and *decay.residual*) were set to constant values and were not explored in this experiment.

Parameter	Value	Parameter	Value
Maxdepth	3	*Maxnodes*	20000
Grow	Yes, Sum of 1, Pext=1.0	*Move*	No
Lruparent	Same	*Phoenix*	No
Local	No	*Decay*	Yes, ? of ?,
			0.5, norou,nores
Deeponly	Symbol	*Addback*	No
Estim	PPM, LinMof, $\lambda = 1$	*Estim.threshold*	Sum of 1

Table 23: Experiment 8: Decay base parameters.

Results: The results of the run are plotted in **Figure 68**. The horizontal axis plots *decay.threshold.thresh* and the vertical axis plots compression. Circles are used for homogenous files and squares for multimodal files. Solid lines denote runs for which *decay.threshold.kind=sum*; dotted lines denote runs for which *decay.threshold.kind=max*. Seven files were tested, making the graph a little messy.

The graph shows that, in general, decaying is detrimental to compression, particularly at low thresholds. The graph carries good news, however, for it shows that if decaying is to be used to keep maximum counts down (so as to simplify arithmetic coding), quite stringent decaying can be used with little loss of compression. It seems for example, that a maximum *sum* of 100 instances could be imposed with little impact on compression. The danger of imposing a low threshold is that it places a bound on the ratio between the most frequent and least frequent symbols in a sample. This bound reduces the compressor's capacity to compress low entropy sources efficiently. However, in most designs there will be little concern for this "best case".

Another solution is to use a *max* threshold which the graph indicates is even more robust; according to the graph, *max* counts can be set as low as 10 with little impact on compression. In fact, this graph shows that for this data, three bit ($\mathbf{Z}[0,7]$) registers could have been used for symbol counts, allowing a 96 byte (but $O(n)$ access) array prediction representation.

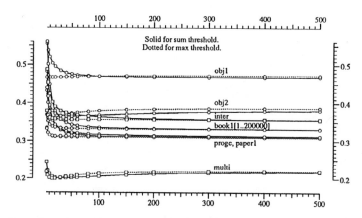

Decaying of samples can be used to introduce contextual local adaptivity and to place an upper bound on frequency counts. However, for ordinary data, decaying has little effect. In this graph, compression (proportion remaining) is plotted against decay threshold (*decay.threshold.thresh*) for a variety of files. Both *sum*(solid) and *max*(dotted) thresholds are used.

Figure 68: Experiment 8: Prop.Rem.
vs *decay.threshold.thresh* (wide angle).

Figure 69 shows a close up view of the interesting zone at $\mathbf{Z}[1, 100]$. Most interesting is that *obj2* exhibits a degree of locality on its *max* curve with a minimum at about 10. Multi has *sum* and *max* optimal values at about 20.

Conclusions: In general, decaying has a detrimental effect on ordinary files. Low thresholds (e.g. < 20 for *sum* and < 10 for *max*) are detrimental to compression. However, for thresholds above 100 *sum* and 20 *max*, decaying has almost no affect on compression. For files generated by moving sources, decaying can provide an advantage ($\approx 4\%$ absolute for these runs). The main benefits of decaying appear to be the reduction of frequency register widths (which simplifies the design of arithmetic coders) and the long term insurance that it provides against sources that move.

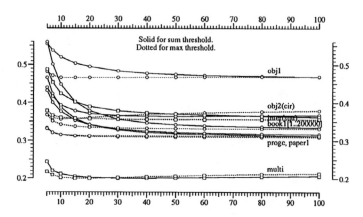

This graph gives a closer view of **Figure 68**. For low thresholds (e.g. < 20), *sum* decaying has a devastating effect on compression, but for higher values appears harmless. *max* decaying can be taken down to 10 without a significant impact. In an optimized implementation, *max* thresholds could be employed to place very low (e.g. 10) upper bounds on frequencies.

Figure 69: Experiment 8: Prop.Rem.
vs *decay.threshold.thresh* (close up).

4.17.12 Experiment 9: Threshold Growth (*grow*)

Aim: To examine the effect on compression of restricting tree growth using frequency thresholding in a compressor that has an initially adaptive structure and asymptotically adaptive contexts.

Method: During the design and implementation of the *grow* and *move* parameters, it was assumed that high depth values (e.g. 5, 10, 20) would be optimal and that nodes would have to be carefully allocated. However, Experiment 5 showed that the optimal depth is 3. This means that for English text files with approximately 27 heavily used symbols, only about $27^3 = 19683$ nodes are likely ever to be needed. Because this value (19683) is less than the number of nodes being used in these experimental runs (20000), the effect of sophisticated tree growth strategies is unlikely to become evident. In this experiment, the memory size is varied (in addition to the growth threshold) so that the advantage of tree management becomes apparent.

In this experiment we are concerned only with the *grow* parameter. The *move* parameter is examined in a later experiment. In this experiment the *move* parameter is turned off. Of the *grow* parameter, only *grow.threshold.kind* and *grow.threshold.thresh* are tested; *grow.probext* is set to 1.0 in this experiment.

The compressor was run over the files *paper1*, *obj1* and *trans* for a variety of memory sizes and growth parameters. The base parameters are listed in **Table 24**.

Parameter	Value	Parameter	Value
Maxdepth	3	*Maxnodes*	?
Grow	Yes, ? of ?, Pext=1.0	*Move*	No
Lruparent	Same	*Phoenix*	No
Local	No	*Decay*	No
Deeponly	Symbol	*Addback*	No
Estim	PPM, LinMof, $\lambda = 1$	*Estim.threshold*	Sum of 1

Table 24: Experiment 9: Threshold growth base parameters.

Results: The results of the run for the *paper1* file are shown in **Figure 70**. The horizontal axis plots *grow.threshold.thresh*. The vertical axis plots compression (proportion remaining). Each line corresponds to a different value of *maxnodes* and is labelled as such.

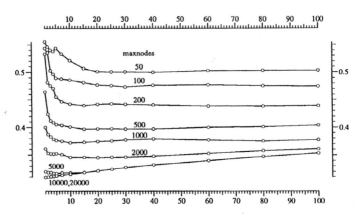

Compressors that never move tree nodes once they has been attached
to the tree must be cautious about where such nodes are placed. This
graph plots compression (proportion remaining) against growth threshold (*grow.threshold.thresh*) for a variety of memory sizes (*maxnodes*)
for the file *paper1*. The point on the far left (*grow.threshold.thresh*=1)
represents the fast growth employed by most algorithms. This is optimal if more than about 2000 nodes are available. For smaller memory,
huge gains in compression can be made by slowing tree growth (an
improvement of 9% absolute for 200 nodes).

Figure 70: Experiment 9: Prop.Rem.
vs *grow.threshold.thresh(sum)*, *paper1*.

This graph shows that if there are less than 1000 nodes, significant
gains can be made by retarding tree growth using an extensibility
threshold (*grow.threshold.thresh*). Taking *grow.threshold.thresh*=10 as
a reasonable production value, the saving in the case of *maxnodes*=200
is about 9% absolute. For 500 nodes it is about 7% absolute. Even for
2000 nodes the technique saves about 2% absolute. For large memory,
compression is lost, but by at most 1% absolute.

At very high thresholds (e.g. 1000 — not shown on this graph) nearly
all the curve sloped upwards. A quick check of **Figure 64** confirmed
that these curves were converging on first order compression.

Of interest is the bump in the 50 curve. This was probably caused by
the increasing sensitivity of the positioning of nodes as the nodes become
scarcer. Other runs with *maxnodes* < 50 were made and the continuity
decreased dramatically as the number of nodes decreased.

The *grow.threshold.kind* parameter was tested by re-running the
experiment with *grow.threshold.kind*=*max*. The results are shown in

Figure 71 which duplicates the previous (*sum*) graph but also includes the *max* runs as the dotted lines. In general, the *max* threshold has little effect, but for low memory it can take off up to 3% absolute.[102]

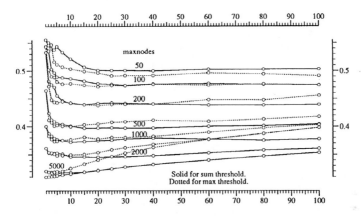

This graph is the same as **Figure 70** except that *max* thresholds have been added as dotted lines. *max* thresholds perform marginally better than *sum* thresholds at low thresholds.

Figure 71: Experiment 9: Prop.Rcm.
vs *grow.threshold.thresh(max)*, *paper1*.

For generality, the experiment was repeated for the files *obj1* (**Figure 72**) and *trans* (**Figure 73**). Short files were chosen because of the increased resource requirements necessitated by the extra dimension (memory).

The *obj1* curves are much sharper than the *paper1* curves. Above the threshold 4, there is little compression to be gained. The curves still exhibit the dramatic drop at the left which shows the usefulness of the threshold. In this case there is an advantage even when there are 10000 nodes. This could be because Markov trees resulting from object files are likely to have a higher furcation than those generated from English text. For 20000 nodes, the cost of using a threshold of 5 (which saves so much for smaller memory) is negligible.

---

[102] Unfortunately, this fact was missed during the experimental process and subsequent experiments use *sum* thresholds rather than *max* thresholds.

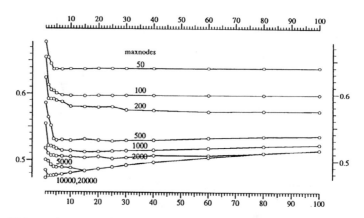

This graph repeats the experiment summarized in **Figure 70** but uses the file _obj1_. The object file exhibits the same dramatic improvement in compression found for _paper1_, with improvements in compression evident even for 10000 nodes.

Figure 72: Experiment 9: Prop.Rem.
vs _grow.threshold.thresh_ for _obj1_.

The terminal session transcript (_trans_) curves are similar to the _paper1_ curves. The same dramatic drop at the left is evident with up to 15% absolute to be gained, 11% if a sensible value (say 10) is chosen for the threshold.

Conclusions: These conclusions refer to the retarding of tree growth through the mechanism of a growth threshold (_grow.threshold.thresh_) in the case of a Markov algorithm that is asymptotically adaptive at the instance level but only initially adaptive structurally and which is compressing a fairly static source. For ordinary files, growth thresholds are very effective for small memory sizes (< 2000 nodes) but have little effect if memory is plentiful. _Sum_ and _max_ thresholds are comparable but _max_ thresholds can yield better compression at low memory sizes at low thresholds. For ordinary files a good value appears to be _sum_ of 10. This yields gains of 5% to 10% absolute for small numbers of nodes while losing at most 1% absolute for large numbers of nodes. If a _max_ threshold is used, the best value appears to be about 5.

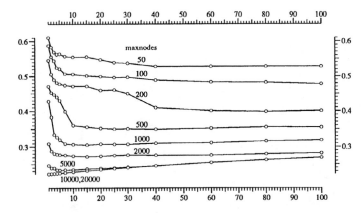

This graph repeats the experiment summarized in **Figure 70** but uses
the file *trans*. Huge improvements in compression are possible here
with gains of 10% absolute for 500 nodes.

Figure 73: Experiment 9: Prop.Rem.
vs *grow.threshold.thresh* for *trans*.

4.17.13   Experiment 10: Probabilistic Growth (*grow.probext*)

Aim: To examine the effect on compression of restricting tree growth, through the mechanism of probabilistic growth, in a compressor that is structurally initially adaptive structure but has asymptotically adaptive contexts.

Method: This experiment is similar to Experiment 9 except that tree growth is restricted probabilistically (*grow.probext*) rather than by a threshold (*grow.threshold*).

The compressor was run over the files *paper1*, *obj1* and *trans* for a variety of memory sizes and probabilities. The base parameters are listed in **Table 25**.

Parameter	Value	Parameter	Value
Maxdepth	3	*Maxnodes*	?
Grow	Yes, Sum of 1, Pext=?	*Move*	No
Lruparent	Same	*Phoenix*	No
Local	No	*Decay*	No
Deeponly	Symbol	*Addback*	No
Estim	PPM, LinMof, $\lambda = 1$	*Estim.threshold*	Sum of 1

Table 25: Experiment 10: Probabilistic growth base parameters.

Results: The results of the run for the *paper1* file are shown in **Figure 74**. The horizontal axis plots *grow.threshold.probext*. The vertical axis plots compression (proportion remaining). Each line corresponds to a different value of *maxnodes* and is labelled as such. In contrast to the threshold graphs, in this graph growth is *slowest* (probability of zero) at the left and *fastest* at the right (probability of one).

This graph shows the same effects exhibited in **Figure 70**. The same massive gains are evident. For a value of *grow.probext* of 0.35, the saving for 200 nodes is about 5% absolute. For 500 nodes it is about 4% absolute. For large numbers of nodes there is a loss of about 1% absolute.

Although probabilistic retardation provides a distinct advantage over rapid tree growth, compression is 40% relative worse than the compression provided by threshold retardation. This is a direct result of stochastic fuzz which ensures some bad growth decisions as well as many good ones. The stochasticity of the method is also evident in the variability of the stability in the curves ranging from smooth at 20000 nodes to jagged at 50 nodes.

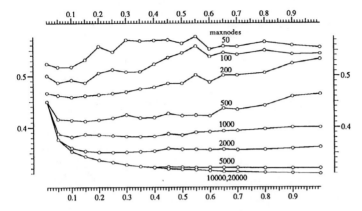

One way of retarding tree growth is to append leaves probabilistically. This graph plots compression (proportion remaining) against the probability of appending a leaf on the end of the matching branch during an update (*grow.probext*). The tree is frozen once all the nodes are placed. Each line corresponds to a different amount of memory. Previous techniques are located at 1.0 at the right of the graph. Probabilistic retardation yields significant gains. However, threshold retardation (**Figure 70**) performs much better.

Figure 74: Experiment 10: Prop.Rem. vs *grow.probext* for *paper1*.

For generality, the experiment was repeated for the files *obj1* (**Figure 75**) and *trans* (**Figure 76**). Short files were chosen because of the large resource requirements necessitated by the extra dimension (memory) in this experiment.

Again we see losses of about 20% to 50% relative to the non-stochastic threshold method.

Conclusions: Probabilistic growth exhibits all the same properties of threshold growth but performs about 40% relative worse because of its stochasticity. The best value for *grow.probext* appears to be 0.4. This setting yields gains (over a probability of 1.0) of about 5% absolute for small numbers of nodes and −1% absolute for large numbers of nodes. The 40% relative degradation over threshold retardation serves as a reminder of the overhead incurred by using stochastic techniques. In this case at least, thresholds should always be used in favour of stochastic growth.

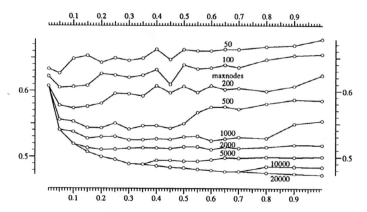

This graph repeats the experiment described in **Figure 74** using the file *obj1*. The same effects are evident.

Figure 75: Experiment 10: Prop.Rem. vs *grow.probext* for *obj1*.

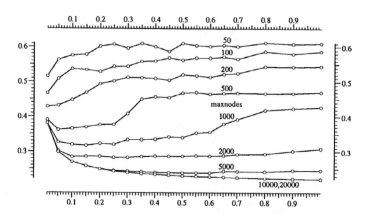

This graph repeats the experiment described in **Figure 74** using the file *trans*. The same effects are evident.

Figure 76: Experiment 10: Prop.Rem. vs *grow.probext* for *trans*.

4.17.14   Experiment 11:   Node Movement (*move*)

Aim: To evaluate the effect on compression of moving leaf nodes within the tree after the supply of nodes (memory) has run out.

Method: Experiment 10 showed that probabilistic growth is inferior to threshold growth and that *sum* thresholds are roughly as good as *max* thresholds for the purpose of tree growth. As a consequence, the only parameter tested here is *move.threshold.thresh*. The compressor was run for a variety of memory sizes and threshold values for the files *paper1*, *obj1* and *objpap*. The base parameters are listed in **Table 26**.

Parameter	Value	Parameter	Value
Maxdepth	3	*Maxnodes*	?
Grow	Yes, Sum of 1, Pext=1.0	*Move*	Yes, Sum of ?, Pext=1.0
Lruparent	Same	*Phoenix*	No
Local	No	*Decay*	No
Deeponly	Symbol	*Addback*	No
Estim	PPM, LinMof, $\lambda = 1$	*Estim.threshold*	Sum of 1

Table 26: Experiment 11: Node movement base parameters.

Results: Figure 77 shows the results for the file *paper1*. The horizontal axis plots *move.threshold.thresh*. The vertical axis plots compression (proportion remaining). The lines correspond to different *maxnodes* and are labelled as such. Solid lines correspond to *grow.threshold.thresh=1* and dotted lines correspond to *grow.threshold.thresh=5*.

As in the earlier experiments, the adaptive mechanism has little effect when memory is plentiful. Here the high memory curves are flat. For less memory, there are significant gains. This graph has the same "look and feel" as the graphs in Experiment 9 but must be interpreted differently. Although the same rising of curves for the low threshold values is present, for this graph the "default" value (i.e. the one used in previous experiments and by other researchers) is *move.threshold.thresh=∞*; the rapid drop at the left indicates only that moving nodes too rapidly is detrimental to compression. Presumably the nodes require time to build up significant samples. More important is the slow drop from infinity back to about 20. This indicates that moving nodes is worthwhile.

The results for the *obj1* file are shown in **Figure 78**. In the light of the previous results, *grow.threshold.thresh=5* was not plotted. The effect of moving nodes is much more pronounced for this run, with the 500-node and 1000-node runs yielding an advantage of about 5% absolute.

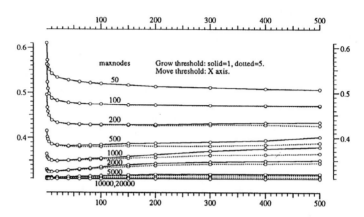

This graph plots compression (proportion remaining) against the parameter *move.threshold.thresh* for the file *paper1*. The lines correspond to different memory sizes. Two different *growth* rates appear: fast(solid) and slow(dotted). Previous algorithms (e.g. PPM) are located at infinity on this graph. The file *paper1* appears fairly stable and the greatest improvement in compression is only about 3% absolute.

Figure 77: Experiment 11: Prop.Rem.
vs *move.threshold.thresh* for *paper1*.

The graph for *objpap* is shown in **Figure 79**. The curves here are smoother, but the same compression advantage is evident — 5% absolute for a memory of about 2000 nodes.

All the graphs indicate that if there are very few nodes (i.e. < 100), it is best not to move the nodes around. This gives the nodes a chance to settle and build up a significant sample.

Conclusions: The policy of relocating nodes in the tree after memory runs out (*grow*) has a negligible effect on compression for large memory ($\approx$ 20000 nodes), a detrimental effect ($\approx$ 3%) for small memory (< 100 nodes) but can produce significant compression gains ($\approx$ 7% absolute) in the mid range ($\mathbf{Z}[100, 10000]$). If nodes are being relocated, there is no advantage in retarding the initial tree growth. Relocation is advised for memory sizes of 100 nodes or greater, for which the best threshold value seems to be *move.threshold.thresh*=20.

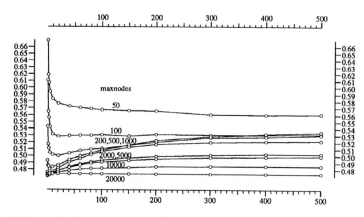

This graph plots compression (proportion remaining) against the parameter *move.threshold.thresh* for the file *obj1*. The lines correspond to different memory sizes. The parameter *grow.threshold.thresh* was set to 1. Previous algorithms (e.g. PPM) are located at infinity on this graph. Node movement yields enormous improvements in compression here — 5% absolute for the 2000 node curve. Again, for small memory, movement is highly detrimental.

Figure 78: Experiment 11: Prop.Rem.
vs *move.threshold.thresh* for *obj1*.

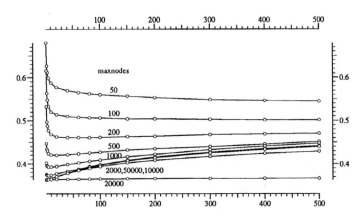

This graph plots compression (proportion remaining) against the parameter *move.threshold.thresh* for the file *objpap*. The lines correspond to different memory sizes. The parameter *grow.threshold.thresh* was set to 1. Previous algorithms (e.g. PPM) are located at infinity on this graph. Node movement yields significant compression gains here — 5% absolute for the 1000 node curve. For small memory, movement is highly detrimental.

Figure 79: Experiment 11: Prop.Rem.
vs *move.threshold.thresh* for *objpap*.

4.17.15 Experiment 12: Four Growth Regimes (*grow,move*)

Aim: To compare the compression performance of four regimes of tree management.

Method: Experiment 9 and Experiment 11 explored the effect of different threshold values for a variety of tree management policies over a variety of files. In this experiment, the best settings of the four distinct policies are run over the three files for a variety of memory sizes. The base parameters for the experiment are listed in **Table 27**.

Parameter	Value	Parameter	Value
Maxdepth	3	*Maxnodes*	?
Grow	Yes, Sum of ?, Pext=1.0	*Move*	Yes, Sum of ?, Pext=1.0
Lruparent	Same	*Phoenix*	No
Local	No	*Decay*	No
Deeponly	Symbol	*Addback*	No
Estim	PPM, LinMof, $\lambda = 1$	*Estim.threshold*	Sum of 1

Table 27: Experiment 12: Growth regimes base parameters.

The four tree management techniques are characterized by their *grow* and *move* thresholds which are listed in **Table 28**. Each technique has been given a two part name depending on the rate at which it grows its tree before and after it has run out of nodes.

Scheme	*grow.threshold.thresh*	*move.threshold.thresh*
fast/stop	1	∞
slow/stop	5	∞
fast/slow	1	20
slow/slow	5	20

This table defines the four tree management schemes used in Experiment 12. *grow.threshold.thresh* determines the rate of tree growth when nodes are still being created. *move.threshold.thresh* determines the rate at which nodes are moved around the tree once the supply of nodes has run out. The four schemes are given names of the form <*growth speed*>/<*move speed*>. Algorithms by other authors can be classified as fast/stop.

Table 28: Experiment 12: Four growth regimes.

Results: The results for *paper1* are shown in **Figure 80**. The horizontal axis plots *maxnodes*. The vertical axis plots compression (proportion remaining). Each line corresponds to one of the four schemes. Scheme *slow/slow* is not plotted in any of these results because it performed almost identically to *fast/slow*. The x-axis only goes to 5000 nodes after which tree management has little effect.

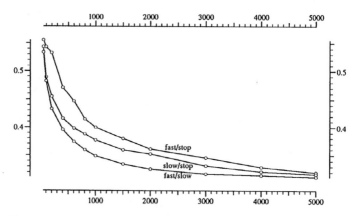

This graph plots compression (proportion remaining) against memory (*maxnodes*) for the file *paper1* for three tree management regimes (**Table 28**). The *fast/slow* scheme performs best, yielding an absolute improvement of 10% over *fast/stop* for 200 nodes.

Figure 80: Experiment 12: Prop.Rem.
vs *maxnodes* for *paper1* (3 regimes).

For *paper1*, the *fast/slow* scheme out-performs the conventional *fast/stop* scheme by about 4% absolute. For 200 nodes, the difference is about 10% absolute.

The results for *obj1* are shown in **Figure 81**. Again *fast/slow* has a 4% absolute advantage over *fast/stop*. The maximum difference here is 12% absolute for 200 nodes. The curves for *obj1* are more jagged than for *paper1*. In particular, the *obj1 fast/slow* curve dips down sooner for small memory.

The results for *objpap* are shown in **Figure 82**. For this file, which changes from object code to English a quarter of the way through, structural adaptivity yields great gains across the entire range of the graph. The difference here is about 14%. For 800 nodes the gain is about 18%. This massive improvement in compression demonstrates the importance of adaptivity in data compression.

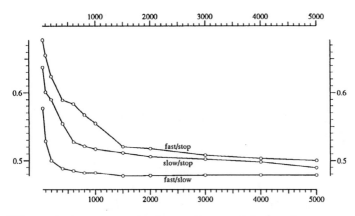

This graph plots compression (proportion remaining) against memory (*maxnodes*) for the file *obj1* for three tree management regimes (**Table 28**). The *fast/slow* technique produces much better compression that the other techniques, yielding an improvement of 12% absolute at 200 nodes.

Figure 81: Experiment 12: Prop.Rem.
vs *maxnodes* for *obj1* (3 regimes).

The book, which is the most homogeneous file, yields the least dramatic results (**Figure 83**). However the technique still yields an advantage, with 5% absolute to be gained at 200 nodes.

Conclusions: As in previous experiments, structural adaptivity is ineffective if memory is plentiful. In this experiment memory was restricted to less than 5000 nodes. The scheme *slow/slow* performed nearly identically to *fast/slow*. The three remaining regimes are ranked from best to worst as follows: *fast/slow, slow/stop, fast/stop*. The technique *fast/slow* is by far the best, beating the other techniques by up to 20% absolute. *slow/stop* enjoys a significant advantage over *fast/stop*. These results show that adaptivity is important in data compression algorithms that use less than 5000 nodes.

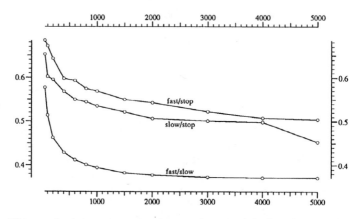

This graph plots compression (proportion remaining) against memory (*maxnodes*) for the file *objpap* for three tree management regimes (**Table 28**). The file *objpap* consists of 20K of object file followed by 50K of English text. In this situation the *fast/slow* scheme yields up to 18% absolute better compression than *fast/stop*. By moving nodes around after they are placed in the tree, the *fast/slow* algorithm adapts to the change in source from object file to English text.

Figure 82: Experiment 12: Prop.Rem.
vs *maxnodes* for *objpap* (3 regimes).

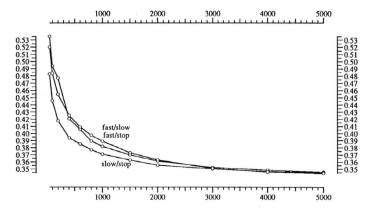

This graph plots compression (proportion remaining) against memory (*maxnodes*) for the first 100000 instances of the file *book1* for three tree management regimes (**Table 28**). Because the book consists of fairly homogeneous English text, the *fast/slow* scheme did not perform as dramatically as it did in **Figure 82**. Nevertheless, it still yielded a compression improvement of 5% absolute for 200 nodes.

Figure 83: Experiment 12: Prop.Rem. vs *maxnodes* for *book1* [1..100000].

4.17.16 Experiment 13: Inheriting Instances (*addback*)

Aim: To determine the effect on compression of adding the sample of a node about to be moved to the sample of the node's parent node.

Method: To test the effect of the *addback* parameter, the compressor was run on a variety of files using the *fast/slow* parameters settings of Experiment 12 (**Table 29**). To ensure that a lot of node movement occurred, only 2000 nodes were used.

Parameter	Value	Parameter	Value
Maxdepth	3	*Maxnodes*	2000
Grow	Yes, Sum of 1, Pext=1.0	*Move*	Yes, Sum of 20, Pext=1.0
Lruparent	Same	*Phoenix*	No
Local	No	*Decay*	No
Deeponly	Symbol	*Addback*	?
Estim	PPM, LinMof, $\lambda = 1$	*Estim.threshold*	Sum of 1

Table 29: Experiment 13: Inheriting instances base parameters.

Results: The results are shown in **Table 30** arranged from best to worst. The third column indicates the effect of addback in absolute proportion removed units.

File	No Addback	Addback	Difference
book1 [1...100000]	0.3627	0.3607	-0.0020
book1	0.3468	0.3466	-0.0002
paper1	0.3257	0.3250	-0.0007
objpap	0.3769	0.3786	+0.0017
obj1	0.4776	0.4796	+0.0020
obj2	0.3497	0.3555	+0.0058
inter	0.3735	0.3797	+0.0062

When a node is moved to another part of the tree, its instances are normally lost. This is potentially detrimental to compression. Loss of instances can be prevented by adding the instances to the parent of the node about to be moved. Unfortunately, in practice the technique does not improve compression, as this table shows.

Table 30: Experiment 13: Addback results.

Conclusions: The effect of adding back is at worst detrimental and at best negligible. It appears to be of no advantage and should not be used.

4.17.17 Experiment 14: LRU Heap Management (*lruparent*)

Aim: To determine the effect on compression of the *lruparent* parameter.

Method: This experiment is identical to Experiment 13 except that the independent variable is *lruparent* rather than *addback*. The parameters for the runs are listed in **Table 31**.

Parameter	Value	Parameter	Value
Maxdepth	3	*Maxnodes*	2000
Grow	Yes, Sum of 1, Pext=1.0	*Move*	Yes, Sum of 20, Pext=1.0
Lruparent	?	*Phoenix*	No
Local	No	*Decay*	No
Deeponly	Symbol	*Addback*	No
Estim	PPM, LinMof, $\lambda = 1$	*Estim.threshold*	Sum of 1

Table 31: Experiment 14: Heap management base parameters.

Results: The results are listed in **Table 32**. The results show that compromising the LRU system by inserting nodes (whose last child node has just been deleted) at the head or tail of the LRU list (instead of in order somewhere in the middle of the list), has *negligible* impact on compression. This result means that the expensive heap management used to maintain LRU order (Section 4.7) is totally unnecessary and can be replaced by a much more efficient (constant time vs logarithmic time) LRU list.

Conclusions: The *lruparent* parameter has a negligible affect on compression. This means that the LRU heap structure described in Section 4.7 could be replaced by a highly efficient linked list structure. For such a list, compression is not affected by whether new leaf nodes are inserted at the head or tail of the list. However, the tail (*lruparent=oldest*) is probably best for the theoretical reasons given in Section 4.7.1.

File	*Oldest*	*Same*	*Youngest*
paper1	0.3262	0.3257	0.3254
obj1	0.4770	0.4776	0.4775
objpap	0.3765	0.3769	0.3764
book1 [1...100000]	0.3627	0.3627	0.3627
book1	0.3468	0.3468	0.3460
obj2	0.3498	0.3497	0.3471
inter	0.3735	0.3735	0.3705

When the last child of a parent node is removed, the parent becomes a leaf node. Because inserting the new leaf in its correct place in a leaf LRU list is expensive ($O(z)$ or at best $O(\log z)$), it is worth investigating other heuristics. This table shows the effect of inserting new parent nodes at the head (*youngest* — least likely to be recycled) or tail (*oldest* — most likely to be recycled) of the LRU list as well as the correct (expensive) position (*same*). The fact that there is no significant difference in the compression performance of the three heuristics means that a fast (constant time *oldest* or *youngest* heuristic) LRU leaf list could be used in practice

Table 32: Experiment 14: Heap management results.

4.17.18 Experiment 15: Tree Reconstruction (*phoenix*)

Aim: To evaluate the compression performance of the strategy of rebuilding the tree when memory runs out (*phoenix*).

Method: This experiment is similar to Experiment 12. However, here when the supply of nodes runs out, the tree is destroyed and rebuilt from the most recent *phoenix.ashes* instances in the history buffer. The base parameters for the experiment are listed in **Table 33**.

Parameter	Value	Parameter	Value
Maxdepth	3	*Maxnodes*	?
Grow	Yes, Sum of ?, Pext=1.0	*Move*	No
Lruparent	Same	*Phoenix*	?, Period=?
Local	No	*Decay*	No
Deeponly	Symbol	*Addback*	No
Estim	PPM, LinMof, $\lambda = 1$	*Estim.threshold*	Sum of 1

Table 33: Experiment 15: Tree reconstruction base parameters.

The compressor was run with these parameters for a variety of files and memory sizes. In each case, the *phoenix.ashes* parameter was set to one quarter of *maxnodes*. This is a good, safe, efficient value for a tree of depth three. Rebuilding the tree with more instances would be likely to result in thrashing. Two values were used for *grow.threshold.thresh*, a *fast* value of 1 and a *slow* value of 5.

Results: The results for *obj1* are shown in **Figure 84**. The horizontal axis plots *maxnodes*. The vertical axis plots compression (proportion remaining). The top two curves (labelled with squares) show the performance of the fast and slow phoenix methods. The lower curve (labelled with circles) is for comparison only and is taken directly from Experiment 12.

Figure 84 shows that slow growth remains effective in the presence of a phoenix mechanism. For 200 nodes, slowing tree growth yields a 5% absolute improvement in compression. However, incremental tree rearrangement remains the best technique. For 200 nodes, incremental tree management yields a 10% absolute improvement in compression over the fast phoenix method.

The results for *objpap* are shown in **Figure 85**. As in previous runs on this file, massive gains in compression are the reward of careful tree management. For 200 nodes, retarding tree growth yields an 8%

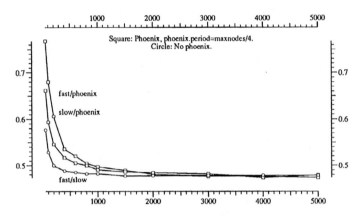

When some Markov data compression techniques run out of memory, they scrap their tree, construct a new tree from the previous *phoenix.ashes* instances and then continue. This graph plots compression (proportion remaining) against memory for the file *obj1* for two such phoenix schemes and the continually adaptive *fast/slow* scheme described in Experiment 12. The *fast/slow* technique yields better compression for small memory sizes.

Figure 84: Experiment 15: Prop.Rem.
vs *maxnodes* for *obj1* with phoenix.

absolute improvement in compression. Full incremental management yields an 18% absolute improvement. Improvements of 7% absolute can be obtained at 1000 nodes.

The same compression improvements are obtained for the book file (**Figure 86**). It is interesting to compare **Figure 86** with **Figure 83**.

Conclusions: If memory is plentiful, the phoenix mechanism will perform nearly as well as incremental tree management. However, as memory decreases, the benefits of incremental tree management increase dramatically. Below 2000 nodes, incremental tree management methods perform between 5% and 20% better than phoenix methods. Retarding tree growth (*grow.threshold.thresh*=5) improves the performance of phoenix methods at these low memory sizes.

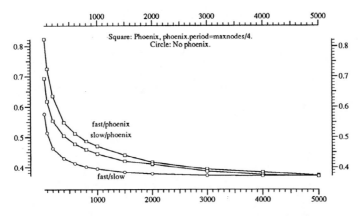

This graph repeats the experiment plotted in **Figure 84** for the file *objpap*. For small memory, the *fast/slow* technique yields far better compression (18% for 200 nodes) than the *fast/phoenix* technique.

Figure 85: Experiment 15: Prop.Rem. vs *maxnodes* for *objpap* with phoenix.

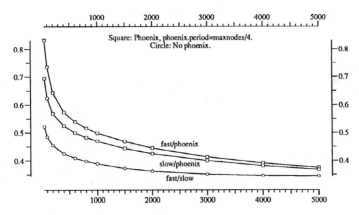

This graph repeats the experiment plotted in **Figure 84** for the first 100000 instances of *book1*. These results indicate that discrete (i.e. phoenix) adaptivity is inferior to continuous adaptivity.

Figure 86: Experiment 15: Prop.Rem. vs *maxnodes*, *book1*[1..100000](+ph).

4.17.19 Experiment 16: Shortcut Pointers (*shortcuts*)

Aim: To evaluate the effectiveness of shortcut pointers (*shortcuts*).

Method: In previous experiments, shortcut pointers were used to increase execution speed. This experiment measures the advantage obtained by using shortcut pointers. The parameters for the experiment are given in **Table 34**.

Parameter	Value	Parameter	Value
Maxdepth	3	*Maxnodes*	?
Grow	Yes, Sum of 1, Pext=1.0	*Move*	Yes, Sum of 20, Pext=1.0
Lruparent	Oldest	*Phoenix*	No
Local	No	*Decay*	No
Deeponly	Symbol	*Addback*	No
Estim	PPM, LinMof, $\lambda = 1$	*Estim.threshold*	Sum of 1

Table 34: Experiment 16: Shortcut base parameters.

The compressor was run with these parameters for two memory sizes and a selection of files. Memory sizes of 200 and 20000 were chosen to represent two extremes of tree volatility.

Results: The use of shortcut pointers had almost no effect on the speed of the compressor, with the speedup for most files being just a few percent. This is probably a result of the complexity and overheads of the SAKDC algorithm. A far better measure of the worth of shortcut pointers is the number of parent to child (*AddLeft*) transitions they avoid.

Table 35 lists the results of the run (rounded to two significant digits). *AvUpdate* is the average of the depths of the deepest node updated for each instance. *AvShort* is the average depth of shortcut pointers taken during the run (this is the average number of hops saved per instance). *RelocRate* is the average number of nodes added to the tree per instance after memory ran out.

In a 20000 node tree, the shortcut pointers take the algorithm most of the way down each branch. When there is a lot of memory, most of the nodes required are in place and node turnover is low. For the 200 node tree, the shortcut pointers reach just over half way down each branch, saving about one hop. The decreased effectiveness of shortcut pointers for smaller trees can be attributed to an increase in node turnover (which increases the rate of invalidation of shortcut pointers) and to the increased sparseness of the tree.

File	*maxnodes*	*AvUpdate*	*AvShort*	*RelocRate*
obj1	200	1.80	1.17	0.27
multi	200	3.00	2.99	0.00
inter	200	2.11	1.30	0.32
objpap	200	1.95	1.05	0.33
book1 [1...100000]	200	2.11	0.93	0.41
obj1	20000	3.00	1.60	0.000
multi	20000	3.00	2.99	0.000
inter	20000	2.90	2.24	0.082
objpap	20000	2.98	2.10	0.040
book1 [1...100000]	20000	3.00	2.41	0.000

Shortcut pointers can be used to avoid most tree traversals as this table shows. Each line in the table corresponds to a single run. *AvUpdate* is the average update depth of the run. *AvShort* is the average depth of traversed shortcut pointers. *RelocRate* is the average number of nodes moved per instance after memory ran out. For large memory (20000 nodes), shortcut pointers span most of the matching branch. For smaller memory, they span about half the matching branch.

Table 35: Experiment 16: Shortcuts results.

Table 36 lists some prediction statistics of general interest. The statistics refer to the samples/predictions in the pivot nodes (Section 4.3.7) during the run.[103] *AvPredDepth* is the average depth of the predictions. *AvPredDis* is the average number of symbols in the prediction. *AvPredSum* is the average number of instances in the predictions. *AvPredMax* is the average maximum frequency in the predictions.

A comparison of **Table 35** and **Table 36** shows that the average prediction depth lags behind the average update depth. This is probably a startup effect as there is no lag for the *multi* run.

The extra 19800 nodes roughly double *AvPredDepth* and halve *AvPredDis*. The *AvPredMax* field, in comparison with the other fields, gives an idea of the spikiness of the samples.

Conclusions: Shortcut pointers are effective at avoiding parent to child (*AddLeft*) transitions. For large memory ($\approx$20000 nodes) they avoid from 70% to 100% of transitions (typically 80%). For small memory ($\approx$200 nodes) in a volatile tree they avoid over 50% of transitions. Whether

---

[103] The statistics refer to the predictions that would actually be made if *estim.merge=DHPC* and *estim.threshold.thresh=1*.

File	*maxnodes*	*AvPredDepth*	*AvPredDis*	*AvPredSum*	*AvPredMax*
obj1	200	1.54	34.5	474	150
multi	200	3.00	4.0	816	370
inter	200	1.79	24.4	1042	207
objpap	200	1.62	23.9	502	78
book1	200	1.70	13.1	509	85
obj1	20000	2.34	13.4	265	232
multi	20000	3.00	4.0	816	370
inter	20000	2.74	8.2	256	191
objpap	20000	2.71	7.9	102	81
book1	20000	2.92	7.8	87	44

This table lists some statistics about the samples of the pivot nodes used for predictions. *AvPredDepth* is the average depth of the pivot node. *AvPredDis* is the average number of symbols of positive frequency in the pivot node. *AvPredSum* is the average number of instances in the pivot node. *AvPredMax* is the average of the maximum frequencies in the pivot nodes. These results sketch a relationship between memory and tree depth as well as giving an idea of the reduction of instance density caused by a deeper tree. Note: "*book1*" is used in this table as a shorthand for *book1* [1...100000].

Table 36: Experiment 16: Prediction statistics.

shortcut pointers will yield an execution speed improvement depends on the particular implementation and the entropy of the source being compressed.

4.17.20 Experiment 17: Final Optimized Benchmark

Aim: To evaluate the strength of the combination of best parameters arrived at in the previous experiments.

Method: In this experiment, the best parameter values of the previous experiments were combined to yield what we call the Opt1 version of SAKDC algorithm (or SAKDC(Opt1) for short). "Opt1" stands for "optimized" version one. In the future, better settings might be found. These can be named "Opt2" and so on.

The Opt1 parameter settings are listed in **Table 37**. The PPMC' algorithm was tested for comparison, as it performed best in the benchmark runs of Experiment 1. The parameters of PPMC' are listed in **Table 38** (a duplicate of **Table 11**, for convenience of comparison). PPMC' uses decaying to provide context adaptivity and the phoenix mechanism to provide structural adaptivity. SAKDC(Opt1) provides no context adaptivity and uses the *move* parameter to provide structural adaptivity.

Despite all the mechanisms tested in these experiments, the final optimal SAKDC(Opt1) algorithm differs from PPMC' in only two major respects: the *grow* parameters and the *decay* parameters. PPMC' uses non-linear estimation and SAKDC(Opt1) uses linear estimation, but this is of little consequence.

Parameter	Value	Parameter	Value
Maxdepth	3	*Maxnodes*	?
Grow	Yes, Sum of 1, Pext=1.0	*Move*	Yes, Sum of 20, Pext=1.0
Lruparent	Oldest	*Phoenix*	No
Local	No	*Decay*	No
Deeponly	Symbol	*Addback*	No
Estim	PPM, LinMof, $\lambda = 1$	*Estim.threshold*	Sum of 1

Table 37: Experiment 17: SAKDC(Opt1) parameters.

Both SAKDC(Opt1) and PPMC' algorithms were run on the entire corpus of files for memory sizes of 20000 nodes and 200 nodes. For the PPMC' 200 node run, *phoenix.ashes* was set to 50.

Results: The results are given in **Table 39**. The columns labelled "P" are for PPMC' and the ones labelled "O" are for SAKDC(Opt1). The other heading numbers are the number of nodes for the run.

Parameter	Value	Parameter	Value
Maxdepth	3	*Maxnodes*	?
Grow	Yes, Sum of 1, Pext=1.0	*Move*	No
Lruparent	Same	*Phoenix*	Yes, Ashes=2048
Local	No	*Decay*	Yes, Sum of 300
			0.5, Rou, Res
Deeponly	Symbol	*Addback*	No
Estim	PPM, NonLinMof, $\lambda = 1$	*Estim.threshold*	Sum of 1

Table 38: Experiment 17: Benchmark parameters for PPMC'.

These results show that for large memory, many of the mechanisms proposed and tested have little to offer. However, for small memory, the mechanisms can produce enormous improvements in compression ($\approx$ 15%). The lack of context adaptivity in SAKDC(Opt1) is evident in its poor relative performance on *artif*.

Conclusions: A compilation of the best parameters from previous experiments reveals that many of the mechanisms proposed are of little use. However, the *move* parameter is worthwhile and can provide huge improvements in compression. In addition, the information obtained from these experiments will allow more efficient Markov compressors to be constructed.

Num	Name	P200	O200	P20000	O20000
1	*bib*	0.755	0.435	0.264	0.263
2	*book1*	0.648	0.453	0.315	0.309
3	*book2*	0.630	0.427	0.288	0.283
4	*geo*	0.700	0.621	0.671	0.590
5	*news*	0.680	0.490	0.350	0.332
6	*obj1*	0.603	0.501	0.497	0.473
7	*obj2*	0.575	0.423	0.348	0.346
8	*paper1*	0.648	0.433	0.311	0.309
9	*paper2*	0.641	0.428	0.307	0.306
10	*paper3*	0.653	0.449	0.338	0.336
11	*paper4*	0.648	0.446	0.366	0.365
12	*paper5*	0.627	0.452	0.381	0.378
13	*paper6*	0.621	0.427	0.318	0.316
14	*pic*	0.134	0.100	0.161	0.102
15	*progc*	0.603	0.413	0.315	0.312
16	*progl*	0.491	0.319	0.239	0.237
17	*progp*	0.511	0.322	0.235	0.230
18	*trans*	0.591	0.390	0.223	0.221
—	average	0.598	0.418	0.329	0.317
101	*artif*	0.215	0.220	0.215	0.220
102	*concat*	0.521	0.379	0.267	0.266
103	*inter*	0.555	0.425	0.368	0.346
104	*objpap*	0.635	0.463	0.374	0.364
—	totaverage	0.577	0.410	0.325	0.314

The results from all the previous experiments were used to arrive at a tuning for SAKDC called SAKDC(Opt1). This table shows the result of this algorithm in competition with the PPMC' algorithm over the corpus of files for two different memory sizes. Compression is expressed as a proportion remaining. The new SAKDC(Opt1) algorithm performs similarly to PPMC' for large memory (20000 nodes) and performs much better (about 18% absolute) for small memory (200 nodes).

Table 39: Experiment 17: Benchmarks
for SAKDC(Opt1) and PPMC'.

4.17.21 Discussion

The experiments presented above are not intended to be exhaustive or conclusive, merely extensive and thorough. They are a measure of progress to date. As well as highlighting the importance of adaptivity in data compression, they provide a general survey that should serve practitioners well. Although the survey is lacking in many respects (e.g. the choice of files), in the absence of any other such survey in the field, the conclusions should rest until challenged by better or broader data.

A few general lessons arose from the experiments.

First, the greatest effect is always produced by the smallest initial force. For example, a little extra memory improves compression dramatically but a lot more improves it only a little. This effect is also evident in other forms such as the need for a history buffer of only 5000 instances to give good compression and the way in which decaying does not impact greatly on compression.

Second, incremental techniques perform better than block techniques, not only because they do not produce sudden, huge delays that are fatal to real time systems, but because they provide steadier performance; the area under a line will always be greater than the area under a saw-tooth curve drawn under the line. This result was evident in the comparison of incremental and phoenix tree management mechanisms. It was also present in the comparison of Markov techniques and Ziv and Lempel techniques (Section 1.13).

Third, there is a remarkable variation in the strength of effect of the different mechanisms. Estimation and blending techniques seem critical whereas instance management seems to have little effect unless extreme.

Finally, exploring a multi-dimensional compressor parameter space can be confusing. It is very easy to be wrong about what will work and what won't.

"In a minute or two the Caterpillar took the hookah out of its mouth, and yawned once or twice, and shook itself. Then it got down off the mushroom, and crawled away into the grass, merely remarking as it went, "One side will make you grow taller, and the other side will make you grow shorter." "One side of *what?* The other side of *what* thought Alice to herself."" *Alice in Wonderland* by Lewis Carroll

4.18 Summary

This chapter has introduced a flexible and powerful variable-order Markov data compression algorithm called SAKDC. The SAKDC algorithm integrates generalized forms of mechanisms found in other algorithms and forms a superset of many previous algorithms such as DHPC and PPM. In addition, the SAKDC algorithm contains a variety of new mechanisms that implement various kinds of adaptivity. These mechanisms include tree growth control, sample decaying, and windowed local adaptivity.

The integration of diverse features produced some problems which have been solved. The problem of maintaining LRU information in a tree has been solved using a new dynamic heap structure. Although made obsolete by experimental results, the technique will find application in other areas of computer science. The problem of maintaining cross-tree links (shortcut pointers) in a dynamically changing tree was identified and solved by allowing suboptimized pointers and by using node incarnation numbers. Windowed local adaptivity was incorporated into all this using a bounded buffer of pointer stacks.

The SAKDC algorithm was implemented and thoroughly tested. Not only did this confirm its design as feasible, but it provided experimental data general to the class of variable-order Markov models. This data could be used to tune production compressors. The conclusions of the experiments are summarized below.

- All the Markov techniques have roughly the same power.

- The best estimation technique is PPM with Moffat estimation (linear or non linear) with $\lambda = 1$. Credibility thresholds do not improve compression unless DHPC estimation is being used.

- The best updating technique is *symbol*.

- A depth of 3 is best in practice. Increasing the depth further does not improve compression.

- Adding memory improves compression until about 10000 nodes, after which extra memory does not improve compression.

- Windowed local adaptivity is generally slightly detrimental to compression but may be of use in keeping down the size of samples for the sake of the coder.

- Decaying is useful for keeping down the size of samples, but otherwise does not improve compression much.

- Mechanisms that introduce structural adaptivity have little effect if nodes are plentiful, but greatly improve compression for small memory ($<$ 5000 nodes).

- Stochastic growth has the same properties as threshold growth but suffers a 40% relative degradation due to its randomness.

- The best growth policy is fast growth while new memory is available, and slow growth when memory has run out and is being recycled.

- Instance inheritance has at best a negligible effect on compression and at worst a detrimental effect.

- Strict LRU management is not necessary, allowing the use of constant time data structures.

- The technique of destroying and rebuilding the tree when memory runs out is inferior to incremental tree management.

This chapter has confirmed the approach to adaptivity outlined in Chapter 3. The introduction of structural adaptivity into an algorithm significantly improves compression when nodes are scarce and the source is moving.

Chapter 5

A Multimodal Algorithm

5.1 Introduction

Of the classes of adaptivity presented in Chapter 3, asymptotic and local adaptivity are of greatest interest. The flexible SAKDC algorithm described in Chapter 4 incorporates both forms of adaptivity as special cases. In this chapter, both locally adaptive and asymptotically adaptive instances of SAKDC are employed in a single algorithm that combines the best aspects of both.

5.2 Combining Models

Many trade-offs are required when designing a data compression model. An example is the trade-off between convergence (asymptotic adaptivity) and tracking speed (local adaptivity). One possibility that has not been adequately explored is that of maintaining many models simultaneously, using only the best-performing model to make the predictions (**Figure 87**). At the expense of extra processing time, the best aspects of a variety of models could be combined without suffering from any of their disadvantages. For example, locally adaptive models and asymptotically adaptive models could be run side by side, forming an algorithm that will converge asymptotically on a fixed source but can also adapt quickly to source movements.

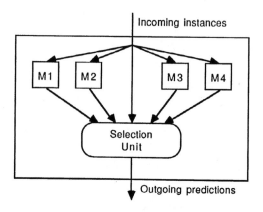

Some of the trade-offs involved in model design can be avoided at the expense of processing time by maintaining many different models simultaneously and using the currently best performing model to make predictions. The selection unit uses the predictions from each of the models along with the incoming instances (centre) to obtain a smoothed performance measure for each model. The entire system is a model in itself and can be slotted into the model unit of **Figure 9**.

Figure 87: Combining models.

The most important component of an algorithm that maintains many models is the mechanism that arbitrates between the models. At each step, the arbitrator chooses one model and uses the model's prediction to code the next instance.[104] The chosen model is the one whose recent performance (as measured by the entropy of recent instances in relation to the model's predictions) is the best.

A trade-off arises in setting the "recency" of the measure. Measurement of the performance of the models over the entire history would prevent rapid transitions between models, whereas measurement based only on the most recent instance of the history would result in chaotic switching.

With the exception of the design of the arbitration unit, the combining of many different models is extremely simple. The *only* costs are the extra memory and extra processing time required, both of which are linear in the number of models.

---

[104] A more general approach is to blend the predictions of the different models. This possibility is not explored in this thesis.

5.3 Multimodal Sources

A compressor that simultaneously maintains a local and an asymptotic model is well suited to compressing both fixed and moving sources. **Figure 88** (which for convenience reproduces **Figure 36** of Section 3.5) shows a number of source trajectories. The reader can verify that a compressor that maintains a locally adaptive model and an asymptotically adaptive model should perform acceptably for all of them.

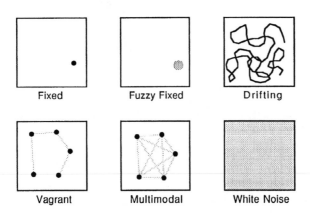

Fixed	Fuzzy Fixed	Drifting

Vagrant	Multimodal	White Noise

This diagram duplicates **Figure 36**. The combination of an asymptotically adaptive model and a locally adaptive model will perform well for a variety of source trajectories. The locally adaptive model will home in on local behaviour (fixed, fuzzy, drifting, vagrant, multimodal) whereas the asymptotically adaptive model will handle long term trends (better convergence on fixed, fuzzy and white noise sources, and conservative predictions for sources with other trajectories that are moving too fast for the local model).

Figure 88: Some interesting source trajectories.

Of particular interest is the multimodal source which jumps between a small, finite number of positions. Multimodal sources are likely to be quite common in practice. In particular, they can be expected wherever files are concatenated together, a prime example being the "source" presented to a communications line. Such a source might alternate between Pascal programs, hex dumps, and object files.

Multimodal sources are interesting because they combine aspects of both fixed and moving sources. Their movement ensures that locally adaptive models will compress them better than asymptotically adaptive models. However, because the source returns to old modes, a locally adaptive model is also inappropriate. In this case, the locally adaptive model's strength of forgetting is also its weakness. As the source jumps between modes, a locally adaptive model will adapt to each mode in turn, forgetting the previous mode.

5.4 Multimodal Models

Better compression could be obtained by maintaining a separate asymptotically adaptive model for each of the source's "modes". Only the model that best models the mode that the source is in would be updated, allowing each model to converge asymptotically on a mode without being contaminated by other modes.

We call the class of algorithms that do this *multimodally adaptive algorithms* or just **multimodal algorithms**.[105]

5.5 Multi-Modal Algorithm Design Issues

Although the idea behind multimodal algorithms is simple, it is by no means clear how such algorithms could be constructed. Not only must the algorithm decide when to switch between models, but it must also decide when a new model should be created. We address these two problems separately.

5.5.1 Model Arbitration

The problem of arbitrating between many models is the same as that of arbitrating between a local and an asymptotic model (Section 5.2) and the solution of maintaining a local performance measure is applicable here as well.

The performance measure of a model should be based on the cumulative entropy of recent instances according to predictions the model made. If $p_{i,j}$ is the prediction of instance j by model i, and K is a locality parameter, then a good performance measure H_i is

$$H_i = -\sum_{j=1}^{K} \ln p_{i,|h|-j+1}\big(h_{|h|-j+1}\big)$$

A disadvantage of this performance measure is that it requires the storage of K entropy values for each model so that the entropy of the K'th

---

[105] The "modal" part of the word "multimodal" is derived from the word "mode" rather than the word "model". Sources have "modes"; compressors have "models".

most recent instance can be subtracted after each new instance arrives. Much the same effect can be achieved using less memory by abandoning the rectangular weighting in favour of an exponential decay.[106] An exponentially decayed measure can be maintained using a single number. Upon the arrival of each instance, the number is multiplied by a decay factor $\rho \in \mathbf{R}(0,1)$ and then the entropy of the new instance is added to it (see also Section 3.7.1).

$$H_i = -\sum_{j=1}^{|h|} \rho^{|h|-j} \ln p_{i,j}(h_j)$$

At each step, the model with the lowest performance measure is used to make the next prediction.

5.5.2 Model Creation

A new model should be created whenever the source jumps to a "new" place, where a "new" place is defined to be a place significantly distant from the old places corresponding to existing models. To detect such new source behaviour, some method is required for determining how far away the source's current position is from the position of each of the models already in existence. A threshold could then be set, below which an old model is used, and above which a new model is created.

There are many ways in which models can be compared. All sorts of tree comparison metrics could be devised. In practice the simplest, most efficient method is to compare the performance of models. A new model could be created whenever all of the old models perform poorly. The difficulty here is defining "poorly"; a new mode might cause a drop in the entropy of all models, hiding the fact that a new model would cause an even greater drop.

A more effective technique is to compare the performance of old models with the performance of a model constructed from the recent behaviour of the source (i.e. a local model). This eliminates the need for subjective interpretations of entropy performance. Whenever the local model out-performs all of the ordinary asymptotic models by a certain factor γ, a new asymptotic model is created, commencing as a copy of the local model. The new model is created only when it seems

---

[106] Another way of saving memory is to store the sum of the entropies of *blocks* of b instances. This would reduce memory usage by a factor of b. The increased granularity would not matter much for largish K. However, blocking would have to be synchronized (or smoothed) for all models so as to avoid saw-tooth effects.

likely that the new model will outperform all other models under some circumstances. These heuristics ensure that a new model is created only if it is a significant distance from the other models. γ is the distance measure.

5.5.3 Memory Allocation

If memory is limited, it must somehow be divided among the models. How this should be done is by no means obvious. It might be advantageous, for example, to let the current model grow at the expense of less used models.

At worst, the best performance might be obtained by allocating all the memory to a single model, in which case the multimodal algorithm provides no advantage.

Justification for splitting memory among a number of models comes from the performance graphs generated in Experiment 6 (Section 4.17.9). Time and time again, we have seen that the first kilogram of compression comes from the first gram of algorithm. For a text file, an algorithm that uses just one node in its Markov tree will remove about 40% of the data. Another 10% absolute can be obtained with an extra 300 nodes. The next 10% absolute requires another 700 nodes. The final 10% absolute (taking the compression to a percentage removed of 70%) requires an additional 9000 nodes (depth 3 *paper1* in **Figure 65** and **Figure 66**). Thus, 10,000 nodes can be allocated to a single model having the power to remove about 70% of the data, or to 5 different models each of which has the power to remove about 65% of the data **(Figure 66)**. The 5% lost in direct performance is made up for in the flexibility of having many models.

In this work, we are more concerned with the mechanism of the algorithm than its resource management. In this study, memory management is simplified by using a fixed number of model slots each of which is allocated a fixed amount of memory. This ensures that the issue of the fairness of arbitrating between models with differing amounts of memory does not arise.

5.6 The MMDC Algorithm

In this section we present the MMDC (Multi-Modal Data Compression) algorithm, which is based upon the discussion above.

5.6.1 Overview

The multimodal algorithm maintains from 1 to *max_models* asymptotically adaptive models (the **ordinary models**) and a single locally adaptive model (the **local model**). Each model generates a prediction for each arriving instance. The predictions are used to maintain a performance measure for each model. The performance measure is the sum of the entropies of the predictions, with exponential decaying (Section 5.5.1) being used to introduce locality. Because the measure is a sum of entropies, a low value indicates good performance and a high value indicates poor performance.

At all times there is an **active model** and a **best model**. The active model is defined to be the ordinary model whose performance measure is lowest. The best model is defined to be the model (ordinary or local) whose performance measure is lowest. At each step, the prediction of the best model is used to code the next instance.

Each arriving instance is used to update only the active model and the local model. Other models use the instance to update their history buffers but do not update (alter) their model (i.e. their tree) at all.[107]

Whenever the local model performs significantly better than the active model (i.e. better than all the ordinary models), a new model is created. If there are already *max_models* models, the least recently used model is destroyed so as to free a slot for the new model. The new model starts as a copy of the local model, but its parameters are set to make it asymptotically adaptive.

Whenever a new model is created, it is put on **trial** for a fixed period called the **trial period**. Only one model can be on trial at a time. New models cannot be created while a model is on trial. If outperformed by an ordinary model, a trial model is instantly destroyed. If at the end of the trial period the local model is performing significantly better than the trial model, the trial model is reset to the local model and commences a new trial period. Otherwise, the trial model is taken off trial, becoming an undistinguished ordinary model.

---

[107] However, the optimization of shortcut pointers can take place because it does not alter the functionality of a model with respect to compression.

5.6.2 Parameters

The parameters of MMDC are mainly to do with timing intervals and performance measures. They are listed below.

max_models: $Z[1, \infty)$. This parameter specifies the maximum number of ordinary models that can be maintained by the algorithm at one time (the number of "model slots"). As in SAKDC, a demand system is used when the slots run out. If a new model must be created when all the slots are full, the least recently used model is destroyed to make room for the new model.

local_model_memory: $Z[1, \infty)$. This parameter specifies the memory of the local model. In this implementation, *local_model_memory* corresponds to the *local.period* parameter of the SAKDC local model.

performance_half_life: $Z[1, \infty)$. This parameter specifies the half-life (in instances) of the negative exponential decay curve of the performance measure (Section 5.5.1). It is the time after which the entropy of an instance is weighted half as heavily as it was when it originally appeared. Because the performance measure consists of a sum of entropies of instances, low values correspond to high performance and high values to low performance.

creation_threshold: $R(0, 1)$. For a new model to be created, the local model must perform significantly better than all the ordinary models. The creation threshold defines "significantly". If the performance of the local model is L and the performance of the best asymptotic model (i.e. the active model) is B, then a new model is created iff $L < B \times creation\_threshold$. The closer this value is to 1.0, the more "trigger happy" the algorithm will be in creating new models.

trial_period: $Z[1, \infty)$. When a model is created, it is in a fairly vulnerable state. This parameter defines the length (in instances) of a safe period, after a model has been created, during which no new models can be created.

SAKDC_local, SAKDC_asymptotic: *record.*   In addition to the parameters of the MMDC algorithm itself, locally adaptive and asympotically adaptive versions of the SAKDC algorithm must be chosen for use as component models.

5.6.3   A Formal Description

The MMDC algorithm is best viewed as a process that reads a stream of instances and writes a stream of predictions (**Figure 89** (similar to **Figure 27** for the DHPC algorithm)). The models are stored in an array called *model*. The local model occupies the position numbered 0 (hence the constant *local*) and the ordinary models occupy the positions numbered $Z[1, models]$ where *models* is the number of ordinary models in existence at a given time. The placement of all the models in a single array allows models to be referenced by index and to be easily enumerated. The array *perf* stores the performance measure for each model.

The *best* and *active* variables contain the index numbers of the best and active models. The variable *trial_rem* is zero if there is no model on trial and positive if there is a model on trial. If positive, the value is the number of instances until the end of the trial period. *trial* is the number of the model on trial and is undefined if *trial_rem*=0. The **any** operator is deterministic, but the actual choice made in the face of equal values does not matter.

The algorithm starts with a single, ordinary model on trial. Starting the model on trial allows the model to be reset if the source commences in a volatile state.

Figure 90 shows the interesting part of the algorithm — the part that performs the model management. This code is executed once per instance. To start, a test is made to see if the trial model, if any, is the best-performing ordinary model (i.e. the active model). If it is not the active model, it is immediately destroyed — if the trial model cannot outperform other active models under the conditions for which it was created then it is unlikely to ever be of much use.

The second **if** statement ages the trial period. If a model under trial reaches the end of the trial period, it must have outperformed all the other ordinary models during that time. This is strong evidence that the source has moved to a new mode. If, however, the local model is outperforming the trial model, it is likely that the trial model was forged during a volatile period of source behaviour and that a brand new model would be more useful. In this case, the trial model is reset to the local model and the trial period restarted in the hope that the source will eventually settle.

The third **if** statement creates a new trial model if there is no trial model and if the local model is performing significantly better than all the ordinary models.

```
process MMDC(in instancestream; out predictionstream) is
    max_models : constant integer ← <parameter>;
    local_model_memory : constant integer ← <parameter>;
    performance_half_life : constant integer ← <parameter>;
    creation_threshold : constant real ← <parameter>;
    trial_period : constant integer ← <parameter>;
    SAKDC_local : constant record ← <parameter>;
    SAKDC_asymptotic : constant record ← <parameter>;
    local : constant integer ← 0;
    best : integer 0...maxmodels ← 1;
    active : integer 1...maxmodels ← 1;
    trial : integer 1...maxmodels ← 1;
    trial_rem : integer 0...trial_period ← trial_period;
    models : integer 1...maxmodels ← 1;
    model : array[0...maxmodels] of SAKDC_model;
    perf : array[0...maxmodels] of real ← 0.0;
begin MMDC
    model[local] ← SAKDC_local(local_model_memory);
    model[1] ← SAKDC_asymptotic;
    loop
        write(predictionstream,model[best].prediction);
        read(instancestream,instance);
        for i in 0...models loop
            perf[i]←perf[i]×decay +
                (− ln model[i].prediction[instance]);
        end loop;
        update(model[local ],instance);
        update(model[active],instance);
        active ← any i: perf[i]≤perf[1...models];
        best ← any i: perf[i]≤perf[0...models];
        <Model Management>
    end loop;
end MMDC;
```

The MMDC main program implements the model unit of **Figure 9**, reading a stream of instances and generating a stream of predictions. The algorithm generates a prediction for each instance before it reads the instance. The algorithm maintains an array of SAKDC models in the variable *model*. The array *perf* holds the cumulative performance measure for each of the models. Only the local model and the active model are updated by each instance.

Figure 89: MMDC main program.

```
<Model Management>=
if trial_rem> 0 and trial ≠ active then
    destroy(model[trial]); trial_rem← 0; dec models;
    <Rearrange models to be contiguous in 1...models>
end if;
if trial_rem> 0 then
    dec trial_rem;
    if trial_rem= 0 then
        if perf[local]<perf[trial]×creation_threshold then
            <Start trial period>
        end if;
    end if;
end if;
if trial_rem= 0 then
    if perf[local]<perf[active]×creation_threshold then
        if models=max_models then
            trial←least_recently_active(models);
            destroy(model[trial]);
        else
            trial←max_models+1;
        end if;
        <Start trial period>
    end if;
end if;
```

This piece of code describes the MMDC model management. The first **if** destroys the trial model (if any) if it is not the active model. The second **if** ages the trial model (if any) and makes it an ordinary model at the end of the trial period; however, if at the end of the trial period, the local model is outperforming the trial model, the trial model is reset to the local model and the trial begins anew. The third **if** creates a new trial model if there is no trial model and the local model is performing significantly better than the active model.

Figure 90: MMDC model management.

Finally, there are two code scraps (**Figure 91**) that were separated out so as to clarify the rest of the code. The first scrap starts a trial period by initializing a trial model to be an asymptotically adaptive copy of the local model. The performance measure is copied over too. The second code scrap is a consequence of the decision to present this code with all models in a contiguous group at the low end of the array. This decision simplified the code, but made model deletion messy, because when a model is deleted, the models above it must be moved down to fill the gap. In addition, the index numbers of the best and active models must be adjusted.

<Start trial period>=
$$model[trial] \leftarrow asymptotic(model[local]);$$
$$perf[trial] \leftarrow perf[local];$$
$$trial\_rem \leftarrow trial\_period;$$

<Rearrange models to be contiguous in 1 ... models >=
$$swap(model[trial], model[models+1]);$$
$$swap(perf[trial], perf[models+1]);$$
$$active \leftarrow \textbf{any } i: perf[i] \leq perf[1 \ldots models];$$
$$best \leftarrow \textbf{any } i: perf[i] \leq perf[0 \ldots models];$$

Starting a trial period involves making an asymptotically adaptive copy of the local model and then setting the timer variable *trial_rem*. The need to rearrange models is a consequence of the decision to store the models contiguously in the array at indices $\mathbf{Z}[1, models]$.

Figure 91: MMDC code scraps.

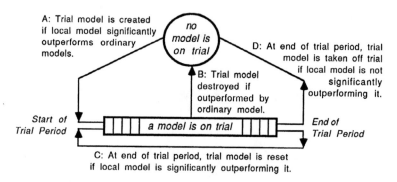

The MMDC algorithm has only two (major) states. These two states correspond to the existence or non-existence of a trial model. There is a time limit on the trial model state which is represented here by a long thin rectangle. The trial model is protected from the local model but vulnerable to ordinary models. At the end of the trial period, if the trial model is performing well, it becomes an ordinary model, otherwise it is reset and the trial period starts again.

Figure 92: State diagram for the MMDC algorithm.

The MMDC algorithm's model management scheme can be summarized by a simple state diagram (**Figure 92**). Each transition is identified by a letter; the letters are not intended to imply an ordering between the transitions. The algorithm can be in one of two states: "trial model" or "no trial model" corresponding to the existence or non-existence of a trial model. The trial model state has a maximum time limit on it (*trial_period*) after which a state transition must occur.

The MMDC algorithm's rather complicated model management scheme arose as a result of experiments with simpler versions of the algorithm. The first version had no trial period; it was assumed that the creation threshold would prevent the local model from significantly out-performing a newly created model, recently its clone. In practice, the creation threshold mechanism proved inadequate.

Without a trial period, models are created in rapid succession when the source becomes unstable. This is most evident during transitions from high entropy sources to low entropy sources. As the local model converges on the new source, its performance measure decreases and it out-performs any model recently cloned from itself, resulting in the creation of another model.

The difficulty with transition periods[108] highlights the tension created by using the creation threshold for two purposes. One the one hand, we wish to detect new modes and create asymptotic models for them as early as possible (i.e. we wish to make the threshold more sensitive). On the other hand, we do not wish to maintain models created during periods of volatile source behaviour (i.e. we wish to make the threshold less sensitive). Use of a trial period prevents models that were created during transition periods from being retained, while allowing the early creation of new models. New models must prove not only that they are better than the other ordinary models, but that they can better the local model as well.

The MMDC algorithm protects the model on trial from the local model while exposing it to attack from other ordinary models. The intention is to give the new model a chance to diverge from the local model while also preparing for the new model's destruction in the event that the source's behaviour isn't new (i.e. if an old model performs better). At the end of the trial period, the new model must have proven useful. If it has not, it is reset in the hope that the source has settled.

---

[108] It may seem strange to talk about a "transition period" for a source that we consider to be switching instantly. The transition period that we refer to is actually an effect of the performance smoothing and model latency.

5.7 Experiments

The algorithm was tested by running it on a variety of artificially constructed multimodal data files. **Table 40** lists the parameters that were used for all the runs. **Table 41** and **Table 42** list the parameters for the locally adaptive and asymptotically adaptive component SAKDC models.

Parameter	Value
max_models	8 models
local_model_memory	1500 instances
performance_half_life	500 instances
creation_threshold	0.9
trial_period	3000 instances
SAKDC_local	see **Table 41**
SAKDC_ordinary	see **Table 42**

The parameters used in the multimodal experiments were arrived at experimentally by tuning the algorithm for the file *multi*. These parameters proved effective in later runs. Early attempts at tuning failed because the *local_model_memory*, *performance_half_life* and *trial_period* parameters were set too low. The *creation_threshold* is important because it determines the entropy "distance" tolerated between models. Here, the creation threshold parameter specifies that a new model must yield 10% relative better compression to establish itself.

Table 40: Multimodal parameters
used in the multimodal experiments.

The local and asymptotic (ordinary) models are identical except for their adaptivity settings.[109] Ordinary models build their tree and then fix its structure. This prevents them from adapting to new sources. To prevent the tree from freezing too quickly, the *grow* extensibility thresholds of ordinary models were set to 2; for a complete branch to be created, it has to receive three instances. In contrast, the local model grows its tree quickly and then uses probabilistic growth (0.7) to sustain its strict local adaptivity. Because there are only 2000 nodes (less than 3 (the depth) × 1500 (the local period)), a perfect locally adaptive tree cannot be guaranteed.

---

[109] Note: At the time of setting the SAKDC parameters for MMDC, only a few of the experiments of Chapter 4 had been performed. Some of the settings may therefore appear inappropriate.

Parameter	Value	Parameter	Value
Maxdepth	3	*Maxnodes*	2000
Grow	Yes, Sum of 1, Pext=1.0	*Move*	Yes, Sum of 1, Pext=0.7
Lruparent	Same	*Phoenix*	No
Local	Yes, Period=1500	*Decay*	No
	=*local_model_memory*		
Deeponly	Symbol	*Addback*	No
Estim	PPM, LinMof, $\lambda = 1$	*Estim.threshold*	Sum of 1

The parameters for the local model were easily chosen with the exception of the *move* parameter which could become applicable during a high-entropy period of the source. To allow continued growth, *move.probext* was set to 0.7. In the light of Experiment 10, a threshold would have been better, but the results of Experiment 10 were unavailable at the time these parameters were set.

Table 41: SAKDC parameters used for the local model.

Parameter	Value	Parameter	Value
Maxdepth	3	*Maxnodes*	2000
Grow	Yes, Sum of 2, Pext=1.0	*Move*	No
Lruparent	Same	*Phoenix*	No
Local	No	*Decay*	No
Deeponly	Symbol	*Addback*	No
Estim	PPM, LinMof, $\lambda = 1$	*Estim.threshold*	Sum of 1

Ideally, ordinary models would be asymptotic, both contextually and structurally. However, making models structurally adaptive also makes them more vulnerable to contamination so in these experiments, the ordinary models were made structurally initially adaptive (*move.active=false*).

Table 42: SAKDC parameters used for the ordinary models.

The parameters listed in **Table 40** were arrived at by experimenting with *multi*, the first of the three data files to be described. At first there was little success because many of the parameters were set too low. In particular, the delay parameters (*local_model_memory* and *performance_half_life*) were set too low to provide reliable performance. The *creation_threshold* parameter also had to be reduced before reliable performance was achieved.

5.7.1 Experiment 18: Artificial Data

Aim: To establish the operation of the multimodal algorithm with data perfectly fitted for it.

Method: Three artificial sources were constructed. Each produced a file of 30000 instances. These three files were then interleaved at 10000 instance intervals producing the final 90000 instance output file which is called *multi* and is numbered 101. This file was fed into the multimodal compressor for this experiment.

Each of the three artificial sources consisted of a solid to depth 3 tree and used only four symbols.[110] Each leaf distribution was chosen by iteratively dividing its prediction probability space. A random symbol was allocated a probability p being a uniformly chosen random number in the range $[0, 1)$. The remaining probability $1 - p$ was divided recursively among the remaining symbols. To ensure that the tree's graph was a strongly connected component, each symbol was given a positive probability not lower than 0.01.

It should be noted that this experiment was not performed in a vacuum, as the *multi* file was used to tune the algorithm's parameters. For a long time the algorithm did not work on this (perfectly suited) data and it was many runs before the copybook graph (**Figure 93**) was generated. However, once found, the parameters served the algorithm well in further runs on real data. In retrospect the main difficulty with tuning the algorithm was that the author had set the *local_model_memory* and the *performance_half_life* far too low.

Results: Figure 93 illustrates the results of the run. The vertical axis measures compression (proportion remaining). The horizontal axis measures time (in instances). The fuzzy vertical lines indicate a change of source mode (referred to here for convenience simply as "sources"). The three sources are named A, B and C and their zones in the data are labelled.

The jaggy black lines of the graph plot the performance measure of each model. The graph was plotted at 50 instance intervals. The performance measure has been scaled to correspond to entropy. The lower the line, the lower the entropy and the better the compression. The black horizontal line at the bottom of the graph is the performance measures of the models available, but not yet created; their performance

---

[110] The fact that a four symbol alphabet was used for these sources had no bearing on the number of symbols assumed by the multimodal compressor which remained at 256.

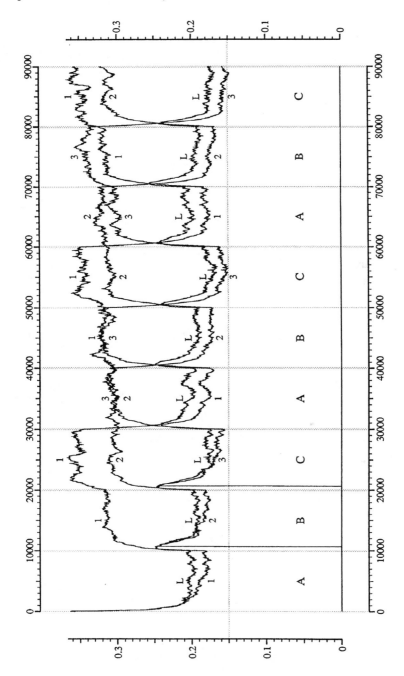

Figure 93: Prop.Rem. vs time for MMDC on artificial data (*multi*).

measure is kept to zero until they are created. When a new model is created, its performance measure is copied from that of the local model. Thus the vertical lines at 10000 and 20000 correspond to model creations.

Each of the models in the multimodal algorithm is numbered. The number that labels each line of the graph is the number of the model that generated the line. The local model is labelled L.

The graph reveals most of the workings of the multimodal algorithm. At the start of the file (at instance one) the source is A and only the local model (model L) and model 1 exist. Model 1 commences on trial and as the active model. Before long, both models adapt to source A; this can be seen from the downward line between 0 and 1500 instances. After about 2000 instances, model 1 diverges downwards from the local model, which remains roughly level. The local model remains roughly level because it bases its predictions only on the most recent 1500 instances and so never builds up samples large enough to converge further. Model 1 diverges because it is continuously accumulating instances from a fixed source.

At instance 10000, the source changes from A to B. Immediately the performance measure of both models increases dramatically. During this period model 1 is contaminated by the instances of source B. If model 1 were exposed to such instances for a long period, it would eventually adapt to them. However, after about 700 more instances, the local model has adapted to source B to the extent that its performance measure is less than 0.9 (the creation threshold) of model 1's (the active model's) performance measure. The result is the creation of model 2 as indicated by the vertical black line rising at 10700 instances. Model 2 starts off as a copy of the local model and with a performance measure identical to the local model. At this point model 2 becomes the active model and model 1 is no longer updated. This prevents model 1 from being further contaminated by source B.

For the next few thousand instances model 2 and the local model joust for the best performance. Eventually (at about 14000 instances) model 2 separates from the local model. This example shows the importance of the trial period; new models must be given a chance to accumulate enough instances to diverge significantly from the local model.

At instance 20000, the source changes to source C and another model is created. The performance of the models in existence rises rapidly and the performance of the local model then falls, causing a new model to be created. The new model, model 3, jousts with the local model but eventually diverges.

At instance 30000, the file switches to source A again. The effect is that the performance measures of all the models rises rapidly. Model 1, however, "recognises" source A and its performance measure falls fast enough to prevent the creation of a new model. Even if a new model were created, it would immediately be cancelled by model 1 (transition B of **Figure 92**). After crossing the local model, model 1 sinks down below the local model and learns some more about source A. One of the advantages of the multimodal algorithm is that models not only don't have to relearn a source but they can pick up more information each time "their" source occurs; the only overhead is the contamination by other sources at the boundaries. This contamination will prevent perfect convergence at infinity.

The fuzzy horizontal line helps to show the result of the interleaved learning. During its first invocation (from 20000 to 30000), model 3 hovers about 2% off the 0.15 mark. During its second invocation (from 50000 to 60000) it hovers a little above the line. During its final invocation (from 80000 to 90000) it hovers close to the line and at one point dips below the line.

The rest of the file is processed in a similar manner, with each model becoming active whenever its source arises. It should be stressed that throughout the run, only the local model and the active model are updated. The rest merely make predictions. Graphically, the best model is the model whose line is lowest at any point of time. The active model is the ordinary model whose line is lowest at a given point in time.

It should be noted that performance measures lag actual performance. For example at 40000 instances, model 2's performance measure takes a long time to drop to 0.17. However, model 2 becomes the best model at about 40500 instances and generates predictions of entropy 0.17 from that point.

The algorithm ran at about 200 instances per second on a Vax8530. This was probably because of the small number of models and the fact that only 4 symbols were ever used.

The actual compression performance of the multimodal algorithm is given in **Table 43**. The performance of four other models is also listed. "Asymptotic" refers to an ordinary model run by itself for the entire run. "Local" refers to the local model. "PPMC'2000" is Moffat's PPMC' algorithm given the same amount of memory as each ordinary and local model (2000 nodes). "PPMC'20000" is Moffat's PPMC' algorithm given the same amount of memory as was allocated to the entire multimodal

Algorithm	Prop. Rem.	Rel. Improv.
Multimodal	0.182	0%
Asymptotic	0.220	17%
Local	0.199	8%
PPMC'2000	0.215	15%
PPMC'20000	0.215	15%

This table lists the compression performance (proportion remaining) of various algorithms on the file *multi*. *Multimodal* and *PPMC'20000* were given 20000 nodes. The other algorithms were given 2000 nodes. *Asymptotic* and *Local* are identical to the component models of the multimodal algorithm. The relative performance is given as the percentage reduction achieved by the multimodal algorithm over each other algorithm.

Table 43: Performance of MMDC on the artificial source (*multi*).

algorithm (20000 nodes). The percentage improvements quoted are *relative* to the performance of the multimodal algorithm.

In this and future experiments, the success of the mechanism of the algorithm will be measured by listing the models that were active during each of a source's activations. For this first run, the results are listed in **Table 44**. This is a perfect performance.

Sources	Active models
A	1 1 1
B	2 2 2
C	3 3 3

This table lists the models that became active during each appearance of the three sources A, B and C in the file *artif*. The MMDC algorithm created a single model for each source and invoked each model only when its source appeared. This is a perfect result

Table 44: Mode detection performance for *multi*.

Conclusions: The multimodal algorithm operated perfectly on specially created artificial data. Exactly one model was created for each source mode and each model became active whenever the corresponding source mode arose. MMDC yielded a 15% relative improvement in compression over PPMC'. This result shows that the MMDC mechanism is basically sound.

5.7.2 Experiment 19: Real Files Concatenated

Aim: To test the multimodal compressor on real data files.

Method: The next stage in testing the multimodal compressor was to run it on real data. Four text files (none in the corpus) were selected for their radically different properties. The files were:

PostScript: A version of the Apple Macintosh LaserWriter header file (length 29005 bytes).

Num: Numbers in ASCII separated by spaces (length 10584 bytes).

HexDump: A hex dump of a VMS .EXE file using the VMS dump command (length 25574 bytes).

Editor: A program written in an editor language (length 27480 bytes).

The resultant concatenated file was concatenated to itself, yielding the test file *concat* numbered 102 in the corpus. This was fed into the multimodal compressor.

Results: A graph similar to that of Experiment 18 was produced and is shown in **Figure 94**. The tick marks on the horizontal axes indicate the points that were sampled when labelling the curves.

In this experiment, the sources were real but the data was repeated. This is evident in the spikiness of the graph and the identical local model performance curves produced by each appearance of the same file.

The graph starts off with the local model and model 1. However, model 1 does not diverge and a new model is created by chance when the local model crosses over at instance 13000. The model is soon destroyed when model 1 out-performs it. The same cycle occurs again at 17000 instances.

At about 30000 instances, source *Num* takes over and model 2 is created. Although model 2 does not diverge from the local model, it must have learned something because it diverges quite well when source *Num* is re-invoked at instance 122000.

A close inspection of the graph reveals that the system is not behaving cleanly. For example, two new models are created during the *Editor* section. What is significant, however, is that no new models are created during the repetitions of the files (except model 6 at instance 185000).

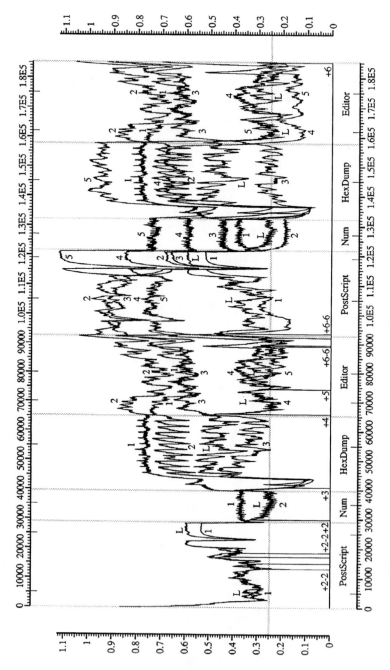

Figure 94: Prop.Rem. vs time for
MMDC on concatenated files (*concat*).

Furthermore, during the repetitions, the active model is separated in general by a large gap from the local model (e.g. during instances 132000 to 157000).

The success of the algorithm in detecting the sources is evident from **Table 45**. Although the *Editor* text provoked the creation of two models, those two models were both invoked at similar intervals when the *Editor* text arose again.

Sources	Active models
PostScript	1 1
Num	2 2
HexDump	3 3
Editor	4,5 4,5

This table lists the models that became active during each appearance of the four sources *PostScript*, *Num*, *HexDump* and *Editor* in the file *concat*. The MMDC algorithm performed perfectly with the exception of the *Editor* source for which two models were created.

Table 45: Mode detection performance for *concat*.

The actual compression performance of the multimodal algorithm is given in **Table 46** in the same format as **Table 43**. Surprisingly, in this run, PPMC' performed slightly better than MMDC. One explanation for this is that in this run, the MMDC algorithm did not use all of the memory available to it. For most of the run, only six models were active (five ordinary models and the local model). This meant that at most only 12000 nodes were being used, rather than the 20000 allocated. A more sophisticated version of the algorithm would avoid this problem by using variable-sized model slots.

Conclusions: The multimodal algorithm performed well on the file of repeated real data. Each source was correctly identified except the *Editor* file to which two models were assigned. This experiment shows that the multimodal mechanism will operate on real data. However, in this run, PPMC' yielded slightly better compression than MMDC.

Algorithm	Prop. Rem.	Rel. Improv.
Multimodal	0.275	0%
Asymptotic	0.377	27%
Local	0.364	24%
PPMC'2000	0.344	20%
PPMC'20000	0.267	-3%

This table lists the compression performance (proportion remaining) of various algorithms on the file *concat*. *Multimodal* and *PPMC'20000* were given 20000 nodes. The other algorithms were given 2000 nodes. *Asymptotic* and *Local* arc identical to the component models of the multimodal algorithm. The relative performance is given as the percentage reduction achieved by the multimodal algorithm over each other algorithm. For this file, the multimodal algorithm performed *worse* than the PPMC' algorithm.

Table 46: Performance of MMDC on concatenated files (*concat*).

5.7.3 Experiment 20: Real Files Interleaved

Aim: To test the multimodal algorithm on real interleaved data.

Method: Experiment 19 tested the multimodal algorithm on highly distinct repeated text files. In this experiment, five (not so distinct) files were selected (four from the corpus, and the *Num* file used in Experiment 19) and merged at random (using a dice) in 10000 instance blocks with the only constraint on the randomness being that no file contribute two consecutive blocks.

Results: The results are shown in **Figure 95**.

The success of the algorithm in picking sources can be judged from **Table 47**. Multiple models were created for *trans* and *progl*. Otherwise, the sources were detected reliably.

Sources	Active models
trans	1 4 1,4 4,5
geo	1 1 1 1
num	2 2 2 3,2
progl	3 3 6 7
book1	5 5 5 5

This table lists the models that became active during each appearance of the five sources *trans*, *geo*, *num*, *progl* and *book1*. The MMDC algorithm had trouble with the sources *trans* and *progl*, but otherwise performed well.

Table 47: Mode detection performance for *inter*.

The actual compression performance of the multimodal algorithm is given in **Table 48** in the same format as **Table 43**.

Conclusions: The multimodal algorithm performed well on real multimodal data (such as might be carried on a network), yielding 4% relative more compression than PPMC'. This experiment confirms MMDC's practical applicability.

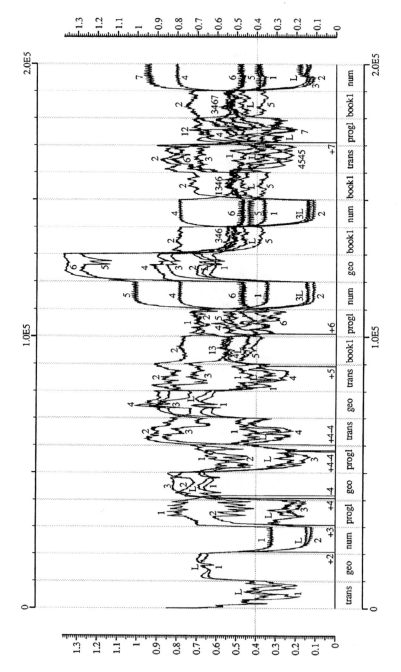

Figure 95: Prop.Rem. vs time for
MMDC on real interleaved data (*inter*).

Algorithm	Prop. Rem.	Rel. Improv.
Multimodal	0.353	0%
Asymptotic	0.405	12%
Local	0.401	12%
PPMC'2000	0.388	9%
PPMC'20000	0.368	4%

This table lists the compression performance (proportion remaining) of various algorithms on the file *inter*. *Multimodal* and *PPMC'20000* were given 20000 nodes. The other algorithms were given 2000 nodes. *Asymptotic* and *Local* are identical to the component models of the multimodal algorithm. The relative performance is given as the percentage reduction achieved by the multimodal algorithm over each other algorithm.

Table 48: Performance of MMDC on real interleaved data (*inter*).

5.7.4   Experiment 21: Effect of Memory

Aim: To determine the effect on the compression performance of MMDC of increasing memory.

Method: Experiment 6 (Section 4.17.9) demonstrated that the returns for using more and more memory in a compression algorithm are diminishing (**Figure 66**). This suggests that the relative improvement of MMDC over PPMC' would increase if more memory were allocated to each algorithm. To test this hypothesis, both algorithms were run on the file *inter* with triple the memory (60000 nodes).

Results: The results are shown in **Table 49**. Although the extra memory improved the performance of both algorithms, MMDC lost ground relative to PPMC' (3% relative in comparison to **Table 48**).

Algorithm	Prop. Rem.	Rel. Improv.
Multimodal	0.348	0%
Asymptotic	0.387	10%
Local	0.401	13%
PPMC'6000	0.376	7%
PPMC'60000	0.353	1.4%

This table lists the compression performance (proportion remaining) of various algorithms on the file *inter*. *Multimodal* and *PPMC'60000* were given 60000 nodes. The other algorithms were given 6000 nodes. *Asymptotic* and *Local* are identical to the component models of the multimodal algorithm. The relative performance is given as the percentage reduction achieved by the multimodal algorithm over each other algorithm.

Table 49: Performance of MMDC with increased memory.

Conclusions: Increasing memory improved the performance of both MMDC and PPMC'. However, the performance of PPMC' relative to MMDC improved. This is a surprising result.

5.7.5 Discussion

The results of the four MMDC experiments are summarized in **Figure 96**. PPMCSmall is PPMC' given the same amount of memory as each submodel of the same MMDC run. PPMCLarge is PPMC' given the same amount of memory allocated to MMDC as a whole. "2E4" and "6E4" refer to the total number of nodes allocated to the each algorithm in each run.

For three out of the four files, MMDC achieved better compression than PPMC', the greatest difference being 15% relative for the artificial data. In general, the difference was small. Nevertheless, MMDC did perform better and this chapter aims only to establish the multimodal algorithm as a new mechanism in data compression. Further research will be required to refine the technique.

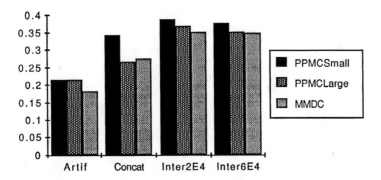

The experiments in this chapter are summarized by this histogram which compares the performance of the MMDC algorithm with the PPMC' algorithm. PPMCSmall refers to PPMC' given as much memory as was given to each component model of MMDC. PPMCLarge refers to PPMC' given as much memory as the entire MMDC algorithm. The two *inter* runs were for a total memory of 20000 nodes and 60000 nodes.

Figure 96: Performance of MMDC
and PPMC' on the multimodal data.

The results of Section 5.7 raise some interesting questions.

• Why did the multimodal algorithm outperform PPMC' by such a wide margin on the artificial data (*mulit*) but not on the real data (*concat*, *inter*)?

- In Experiment 20 (Section 5.7.3), why did the multimodal algorithm fail to generate a new model at instance 10000 (**Figure 95**), a point at which a radically different source took over? Why did model 1 subsequently perform well?

- Why did the multimodal algorithm perform worse than PPMC' on the repeated files (*concat*)?

- Why did the multimodal algorithm's compression relative to PPMC' decrease when the memory size was increased?

The following explanation has not been experimentally verified but fits all four phenomena remarkably well.

Consider a Markov tree model that is compressing a file consisting of three segments generated by three sources X, Y and Z of depth 2. Source X generates symbols a and b according to its (depth 2) conditional probability distributions. Source Y generates the same symbols but according to a quite different probability distribution. Source Z generates the symbols c and d according to yet another probability distribution.

Now consider the effect of the sources on the multimodal algorithm (**Figure 97**). During source X's segment, model 1 grows six nodes and accumulates instances in them. By the time source Y arrives, model 1 has accumulated many instances and is well adapted to source X. The new instances from the radically different source Y hardly impact on model 1's samples (which contain many instances) and so model 1 starts performing poorly. The local model performs better and model 2 is created. During source Y's segment, model 1 is not updated. Model 2 grows a tree identical to model 1's tree except that its samples are entirely different, reflecting the different transition probabilities of source Y.

When source Z arrives, the input consists of instances of the symbols c and d. As no part of model 2's tree deals with these symbols, six new nodes (a new subtree in fact) are very quickly grown. Because the new subtree contains no instances, it adapts *as fast as the local model* to the new source Z. The result is that the local model never out-performs model 2 and no new model is created.

Central to this explanation is the concept of **tree zones**. A source mode's **zone** is a subset of nodes in a tree that the source mode generally "inhabits". Two source modes are said to **collide** if their zones overlap. So long as there are no collisions between source modes, there is no advantage in using different trees to model them, as a single tree can contain all the models without the models interfering with each other.

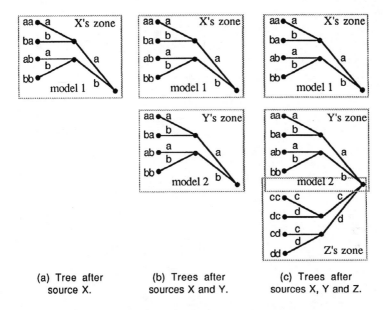

(a) Tree after
source X.

(b) Trees after
sources X and Y.

(c) Trees after
sources X, Y and Z.

This diagram illustrates how the tree zones of different source modes could affect the operation of the MMDC algorithm. These pictures depict the state of two models after data from various sources has arrived. In (a), a source X has caused model 1 to be created. In (b), source Y has appeared. Because it uses the same zone as source X, it collides with source X causing the creation of a new model. In (c), source Z has appeared but because its tree zone does not collide with that of source Y, it is incorporated into model 2 as a distinct zone and no new model is created.

Figure 97: The tree zone explanation of MMDC's performance.

The concepts of zones and collisions provide answers to the questions posed earlier.

Conjecture: The multimodal algorithm performed significantly better (relative to PPMC) on the artificial data than on the real data because the artificial data was generated by sources whose trees were solid to depth 3. The three artificial sources collided (overlapped) completely.

Conjecture: The multimodal algorithm failed to create a new source at instance 10000 in Experiment 20 because the two sources (on either side of 10000) occupied different tree zones. The first source section was a terminal transcript (ASCII data) and the second was geological data (containing many null bytes and low numbered bytes).

If two source modes are similar, a collision can actually *improve* compression because the overlapping models will improve each other's statistics. This is what happens in an ordinary Markov model.

Conjecture: The multimodal algorithm performed worse than PPMC' on the repeated files because the files that were radically different occupied different tree zones and the sources that were similar fed off each other's statistics.

Another factor is memory. In the zone example, model 2 had enough memory to allow it to create the c, d subtree for mode Z. If it had not had the memory, the subtree would not have been created, the model's performance would have dropped, and a new model would have been created. Thus memory is a factor in model creation.

Conjecture: The multimodal algorithm's performance *relative* to PPMC' decreased when more memory was added because with more memory, new modes were more likely to grow in old models.

A closer examination of the results of Experiment 21 revealed that only five 6000 node models had been created; seven 2000 node models were created in Experiment 20. The problem of ordinary models adapting too quickly to new modes was anticipated early in the experimental process. That is why the *grow.active* parameter of the ordinary models was set to *false*.

It may appear from this discussion that the multimodal algorithm can never yield an improvement, as sources that don't collide can have no effect on each other and sources that do collide tend to enhance each other's performance. However, in a third case of two sources that collide *and* have quite different statistics, the multimodal algorithm will excel, as it did in Experiment 18. In practice, the situation is likely to be much messier with most sources partially colliding.

5.8 Other Issues

MMDC is a completely new kind of data compression algorithm. Until now, research has been oriented towards the development of more and more sophisticated locally adaptive algorithms. MMDC steps sideways by addressing issues of **model management**. There is much to discuss.

5.8.1 MMDC is Really a Meta-Algorithm

Although our present implementation of MMDC uses the SAKDC algorithm described in Chapter 4, the MMDC algorithm is not strictly bound to it. All that the MMDC algorithm requires is locally adaptive and asymptotically adaptive versions of a component model. MMDC is not even bound to the modern paradigm — there is no need for the input to be processed one instance at a time, nor for there to be a common, shared coder. All that is required is that it be possible to measure the performance of each model for arbitration purposes.

In Section 3.4 we saw that most data compression algorithms can be modified to be non-adaptive, initially adaptive, locally adaptive or asymptotically adaptive. There is no reason to believe that any algorithm could not be used as a component model in the MMDC algorithm.

Thus the MMDC algorithm is really a **meta-algorithm**, an algorithm for manipulating and choosing other algorithms.[111]

An exciting prospect is that of building multimodal algorithms from LZ models (Section 1.7). LZ models do not compress as well as Markov models, but they run much faster.[112] A multimodal algorithm using LZ models is likely to yield better compression than ordinary LZ algorithms, while still remaining fast enough to be practical.

5.8.2 Efficiency

MMDC maintains many models simultaneously. This can be expensive in CPU time. Fortunately, only the local model and the active model are actually updated; the other models must generate predictions but are not updated.[113] Generating a prediction involves shifting the history buffer and taking a shortcut pointer to the next matching branch.

Maintaining a performance measure for each model is expensive because the logarithm of the probability of the symbol of each arriving instance has to be calculated and summed. This operation could be sped up with the help of a logarithm lookup table. Another solution is to use the new fast arithmetic codes and simply measure the number of output bits.

---

[111] The word "algorithm" in this section could just as well be replaced by the word "model".

[112] Moffat[Moffat88] reports that a highly optimized LZ algorithm ran about eight times as fast as a highly optimized version of the PPM algorithm.

[113] However, shortcut pointers can be optimized as they do not affect the functionality of a model, only its speed.

Maintaining identification of the least recently active model is not the problem that it was in Chapter 4. Here the "things" are not coloured red and green and there are fewer of them. A standard doubly linked list can be used.

Copying the local model is another expensive operation. The operation could be avoided by altering the algorithm so that whenever a trial model is required, the local model is converted to an asymptotic model (becoming the trial model) and a new empty local model is created from scratch. Another solution, assuming the copy operation must take place, is to use only *relative* pointers in models. This would allow models to be copied by a single memory copy instruction rather than by recursive structure traversal.

In this implementation, copies of local models were constructed by feeding the local model's history buffer into a newly created model. This was not fast, but it was simple and reliable.

Multimodal data compression is perfectly suited for parallel execution. The component models, the arbitration unit and the coder could be placed on separate processors. Because the models do not communicate with each other, interprocess communication would be minimal. Although a straightforward parallel implementation[114] would require that the models be synchronized at the instance level, appropriate buffering would render this unnecessary.

5.8.3 The Use of Non-Adaptive Models

In addition to managing a set of asymptotic models, a multimodal algorithm could run a group of non-adaptive models tuned to particular commonly occurring sources. The number and nature of models used is bounded only by the deviousness and ingenuity of the model constructors. On a highly parallel machine, inclusion of extra static models would cost nothing in time except the slight increase in the $O(\log k)$ blending time needed to compare the performance of k models.

5.8.4 Heterogeneous MMDC

Although powerful, Markov models perform poorly on some sources. Signal data, for example, is better compressed using linear prediction [Witten80]. To cater for different classes of data, more than one class of model could be manipulated by the MMDC algorithm.

---

[114] Some researchers might argue that this is a contradiction in terms.

Each class would have its own local model. However, as in homogeneous MMDC, only one model from the entire set of models would be updated.

The combination of multiple classes of asymptotically adaptive model along with a suite of non-adaptive models would result in an extremely versatile algorithm capable of efficiently compressing messages generated by a variety of interleaved sources. This flexibility is bought at the cost of an increase in processing time.

5.8.5 MMDC as Unifier

The field of data compression was launched by Huffman coding and was carried by ad-hoc techniques for many years. The emergence of the modern paradigm of data compression provided a much purer view of the field, in which a single model supplies predictions through a narrow interface to an arithmetic coder. With the advent of multimodal data compression comes the possibility of combining many different kinds of model. If processors are cheap, multimodal techniques will encourage the inclusion of any model that is vaguely orthogonal to the others. We are likely to find ourselves back where we started: in a maze of ad-hocery.

Whatever happens, we can at least be assured that the ad-hocery will be contained. An advantage of the modern paradigm and the MMDC algorithm is that they provide strong frameworks within which models must reside. No matter how different the models may be on the inside, their interfaces must all be identical; all must accept instances and produce predictions (or perhaps generate code). Multimodal data compression integrates nearly all other data compression techniques by providing a framework within which different models can reside without interfering with each other.

5.8.6 A Note on Security

Before finishing this chapter, it is worth pointing out a rather surprising security implication of multimodal algorithms. A multimodal algorithm set up on a public communications channel will store information about messages produced by various "sources" that have used the channel in the past. If users of the channel are able to measure the compression yielded by the channel (for example by measuring the time that the channel takes to transmit the message), they will be able to obtain information about messages that have previously been transmitted. This problem exists with adaptive data compression but is far more severe for multimodal algorithms which are capable of "recognising" a document that was transmitted a long time before.

For example, if a multimodal compressor were connected to a user's terminal line (so as to reduce the transmission time), an intruder who had gained access to a user's line could tell if the user has been scrolling particular files simply by scrolling them himself and seeing how fast they appear. More tenuous channels are apparently already in use in the intelligence world[**Wright87**].

The problem cannot be alleviated by cryptography, as data compression must be performed *before* any cypher layer takes control of the data. The only solutions seem to be to hide the compression performance or to isolate the compression data streams of different users.

Hiding compression performance may be possible over a multi-user line by delaying all messages at the receiving end so as to provide a fixed (worst-case) transmission speed. From each user's point of view, the channel has a constant speed and no data compression is taking place. From the channel's point of view, the (multi-modal) data compression is reducing traffic volume, allowing more (fixed rate) users to use the channel.

Isolation of users' compression data streams can be accomplished by maintaining a separate multimodal model for each user or by restarting the data compression system whenever the channel user changes. The former scheme loses the simplicity of centralization and the later scheme loses some of the advantages of multimodal models.

5.9 Summary

This chapter introduced the technique of dynamically swapping between different models (e.g. locally and asymptotically adaptive models) so as to maximize the benefits and minimize the disadvantages of each. From this grew a technique for the compression of multimodal sources, in which asymptotic models are created dynamically, one for each of the source's modes. New models are created whenever a locally adaptive model out-performs all the asymptotically adaptive models. This method was used to construct the MMDC algorithm which has been described in detail. Experiments showed that the MMDC algorithm reliably detects source modes, creating a new model only when a new source mode appears. MMDC is really a meta-algorithm because it can use any component model so long as the component model exists in locally adaptive and asymptotically adaptive forms. MMDC is ideally suited for parallel execution because its models are loosely coupled. Extensions for different classes of models and a library of fixed models make MMDC an ideal base for the combination of many different data compression algorithms. Multimodal algorithms introduce new security problems because of their tendency to store sophisticated models for long periods of time.

The development of the MMDC algorithm is a good example of how science builds upon itself. It is hard to imagine the MMDC algorithm arising without some concept of model or without a clear categorization of adaptivity, but neither of these concepts were developed with multimodal adaptivity in mind.

Chapter 6
Applications to User Interfaces

6.1 Introduction

Data compression techniques are used to reduce the volume of data being conveyed through a channel. Applications of data compression are distinguished by the nature of their channel. For data transfer, the channel is a communications line. For data storage, the channel is a storage medium. Other applications, such as authorship identification [Roberts82], which rely on fluctuations in data compression performance, use an imaginary channel. In this chapter we introduce a new applications area whose channel is the interface between a user and a computer terminal.

6.2 A New Application

The motivation for the new application is that the behaviour of users can be predicted just as the behaviour of files can be predicted. By presenting these predictions to the user in a useful form, they can be used to reduce the amount of work the user has to perform.

A user prediction system might operate as follows. The user works at the terminal entering commands and receiving responses on the screen. Inside the computer there is a special process, independent from the user's process, that records everything that the user types (i.e. the user's input stream). The process uses the data compression techniques described in this thesis to predict what the user is going to type next. If the program has confidence in its prediction, it displays the prediction in a special place on the screen. The prediction consists of a string of characters.[115] This contrasts with our previous view of predictions as probability distributions. The user observes the prediction, and if it is the same as what was about to be typed, the user hits a specially reserved key which enters the predicted string as if the user had typed the string directly.

The effect of all this is to reduce the number of keystrokes that the user needs to make. At each step, the system programs the special key with the characters that it thinks the user is likely to type next.

6.3   A Paradigm of User Prediction

The new application deviates from the modern paradigm of data compression. Nevertheless, the major components can still be identified. The user is both source and transmitter. The terminal keyboard is the channel. The model and decoder exist in software.

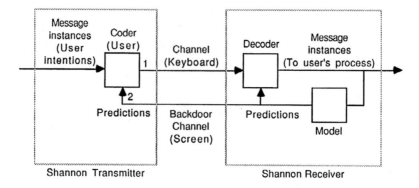

The modern paradigm of data compression can be modified for application to user interfaces. In the modified paradigm, the receiver generates predictions which are presented to the user. The user has the option of typing normally or confirming one or more predictions by pressing special keys. Predictions must be multi-instance in order to save keystrokes.

Figure 98: The modern paradigm modified for user interfaces.

---

[115] Up until this point, we have referred to symbols as "symbols". In the context of user interfaces in which the user nearly always types ASCII characters, we can take the liberty of referring more specifically to "characters".

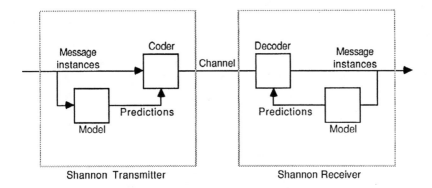

This figure is a duplicate of **Figure 9**, placed here for convenient comparison with **Figure 98**.

Figure 99: The modern paradigm of data compression.

Figure 99 shows the modern paradigm of data compression. **Figure 98** shows the modern paradigm modified for user prediction. Predictions are displayed on the user's screen at position 2. The user types keys on the keyboard at position 1. The receiver has not much changed but the transmitter differs substantially. Instead of having a coder and predictor, there is only a coder. Predictions originate from the receiver.

It is unusual for information to flow backwards in a data compression system. The aim is usually to reduce the traffic between the sender and the receiver. In this case, the cost of sending from the receiver to the sender is negligible compared to that of sending from the sender to the receiver, and so it is possible for all the predicting to be done by the receiving end. Thus, the receiver *presents* the alternatives and the sender *selects* from them.

6.4 Examples

We now present some practical examples of how a user predictor might be used. It is not intended that a user predictor produce helpful predictions at all times. The system would be useful even if it produced predictions that saved the user just 5% of typing. For some data, the success rate will be much higher.[116] A prediction system would be useful in reducing user frustration even if it were only required on the occasions where the input required was tiresomely repetitive.

---

[116] For example if the user were typing the words of the song "The Twelve Days of Christmas".

6.4.1 Debugger Sessions

A debugger program is being used to debug a program. At first, the user uses the debugger only for simple checks. However, as the program grows bigger and the debugging task grows more complicated, so does the number of "setting up" commands that need to be executed whenever the debugger is entered.

Example setting up commands might be (VAX/VMS Debugger):

```
SET MODE DISPLAY
SET LANGUAGE PASCAL
SCROLL/DOWN:145
SET BREAK/LINE/AFTER:1000
GO
```

After one or two debug sessions with this setting up sequence, the prediction system acquires enough information to predict the sequence. The next time the user enters the debugger and types debug$^\wedge$Mset,[117] the prediction system predicts the rest.

Although the user could have anticipated the repetition and placed the commands in a command file, in practice the user is more likely not to notice the build up of commands and type them each time. Eventually the user will recognise the problem and write a command file. A prediction system could help create such a file.

6.4.2 Typist

A typist is typing in a document in which the word "indistinguishable" occurs very often. After the typist has typed it a few times, the prediction system catches on to the string and starts predicting it after (say) "indi". Eventually, the typist types the word as "indi"▷, (where ▷ is a single key that enters the prediction system's current prediction), thus saving thirteen keystrokes per occurrence.

That sufficient repetition occurs in manuscripts to warrant a prediction system is indicated in the following discussion of the use of the macro facility in the TEX typesetting system[Knuth79][Knuth84]. Knuth has mathematical TEXts in mind but the quote does illustrate the problems and trade-offs involved in reducing repetition.

> "Of course, you usually won't be making a definition just to speed up the typing of one isolated formula; that doesn't gain anything, because time

---

[117] The notation "$^\wedge$*letter*" will be used to represent a control character. In this example $^\wedge$M is the ASCII "RETURN" character (code–13) used to terminate lines of input.

goes by when you're deciding whether or not to make a definition, and when you're typing the definition itself. The real payoff comes when some cluster of symbols is used dozens of times throughout a manuscript. A wise typist will look through a document before typing anything, thereby getting a feel for what sort of problems will arise and what sort of definitions will be helpful." [Knuth84](chapter 20).

As well as helping to replace the simple use of macros, a prediction system can be helpful at the times when the overheads associated with using other methods (e.g. definitions) are prohibitive.

6.4.3 Indenting a Program

A programmer is editing a program and needs to indent a procedure by three characters per line. The programmer starts with "⊔⊔⊔^M"[118] and repeats it. After a few repetitions, the prediction system catches on and the programmer uses a single keystroke for each repetition.

6.5 Review of Interactive Environments

The user predictor proposed in earlier sections has the potential to become an important and useful part of interactive environments. However, a multitude of user-environment tools have already been developed. Because most of these tools have been developed for programmers, the field is usually referred to as the field of **Interactive Programming Environments** (IPEs for short). The aim of this section is to position user predictors in this field.

6.5.1 Inner and Outer Environments

Barstow and Shrobe[**Barstow84**](section 2.4) have suggested that in future IPEs, programs under development will be tightly bound with the tools that manipulate them. At present, program and tools can still be distinguished. We divide environments into two parts: an *inner environment* and an *outer environment*.

The **inner environment** consists of the object that the user is working on and the set of primitive operations used to manipulate it. In a programming environment, the object would be a computer program, and the primitives would be the commands to edit and run it. The program and the set of primitives can be considered formally as being an instance of an abstract data type[Guttag80].

---

[118] "⊔" will be used on occasions to represent the space character.

The **outer environment** consists of a set of sophisticated tools whose purpose is to amplify the user's actions. These tools are based upon (but are not necessarily symbiotic with) the tools of the inner environment. Currently, few environments have an outer environment. Examples of outer environments are the Programmer's Assistant[**Teitelman72**][**Teitelman84**], the Programmer's Apprentice[**Rich78**][**Waters82**], DWIM[119] Spelling Correction[**Teitelman72**][Teitelman72] and Active Help Systems[**Fischer84**]. Although all these tools can ultimately be considered to be an extension of the abstract data type of the inner environment, this may not be a very useful view to take.

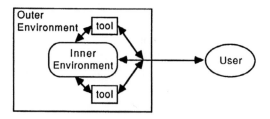

Modern environments can be divided into an *inner environment* and an *outer environment*. The inner environment consists of the objects on which the user is working, along with tools to manipulate the objects. The outer environment looks in to the inner environment (as the user does) and attempts to aid the user by providing powerful facilities for issuing commands to the inner environment. Examples of outer environments are the Programmer's Assistant, the Programmer's Apprentice, DWIM and Active Help Systems.

Figure 100: Inner and outer environments.

The relationship between the user, the outer environment and the inner environment is confusing because the outer layer can vary its thickness. The best way to imagine the system is to consider the user to be communicating with the inner environment with the outer environment intercepting the communication from time to time (**Figure 100**). At times, the user directly manipulates the inner environment, whereas at others all communication is channelled through the complex mechanisms

---

[119] "DWIM *(dwim) noun.* A complicated procedure (in the INTERLISP dialect of LISP) that attempts to correct your mistakes automatically. For example, if you spell something wrong or don't balance your parentheses properly, it tries to figure out what you meant. DWIM stands for "Do What I Mean". When this works, it is very impressive. When it doesn't work, anything can happen. When a program has become very big and complicated — so complicated that no one can understand how to use it — it is often suggested in jest that dwim be added to it."[**Steele83**](p. 59)

of the outer environment. The outer environment should not distract the attention of the user from the inner environment unless it can help in a significant way. To summarize, the inner environment defines the domain in which the user is working, and the outer environment provides entities that enhance the interface to the domain.

Tools in the outer environment exist to serve the user. Unlike objects in the inner environment, they do not define the domain in which the user is working but rather "look in", as the user does, to the inner domain. Commands issued to the outer environment are likely to be less specific than those to the inner environment. As a result, outer environments often contain user models which are used to resolve ambiguity and choose a course of action.

Having established a split-level view of the field of IPEs, we turn our attention to previous work in interactive systems design and other areas that relate to automatic prediction.

Research into IPEs that is relevant to user prediction can be sorted into roughly the following categories.

- General design of interactive systems.

- Programmer's helpers.

- The design of command languages.

- Modelling the user at the keystroke level.

- Modelling the user at the command level.

The work in each of these categories will be discussed in turn.

6.5.2 General Design of Interactive Systems

There is no shortage of work on interactive design. Interactive design has recently become a popular research topic because of the increasing cost of human resources. This work can be split into two groups. The first group typically concerns itself with low level terminal interface design (e.g. **[Morland83]**(4.2)). Schneiderman**[Schneiderman82]** discusses command languages, response time, the wording of system messages, help facilities and other issues before getting on to his main point which is that interactive systems are important and can be made easy to learn and use by employing the principle of direct manipulation. The principle of direct manipulation requires that a model of the world being manipulated be built, and that it be directly and simply manipulated by the user. The second group concerns itself with the implementation of such systems and methodologies for their development[**Freeman78**] [**Zunde81**][**Wasserman82**](p. 6).

None of this work is particularly relevant to prediction systems as it mainly deals with the design of conventional environments.

6.5.3 Programmer's Helpers

Programmer's helpers are outer environment programming tools.

Researchers on the Programmer's Apprentice (PA)[Rich78][Waters82] describe it as "midway between an improved programming methodology and an automatic programming system." Fundamentally it is a high level programming language cast in interactive form.

The programmer's apprentice is a good example of an outer environment component. The following extract gives a very good feel for the PA and the nature of inner and outer environments.

> "The programmer is the active agent in the picture. He issues commands directing the components in the environment in order to create and modify programs. The programmer's apprentice system (PA) is an additional active agent which assists the programmer with the task of programming. There are three important points about the way the PA fits into this picture. First, it is not intended to replace the programming environment, but rather to augment it. It will communicate with the rest of the environment in terms of code, and commands. Second, the programmer can still communicate directly with the rest of the environment. This gives him a trap door so that he is not always required to work through the PA." [Waters82]

The programmer's apprentice differs in many ways from a user predictor. It is much more intelligent than a user predictor and contains more knowledge about the inner environment.

> "The PA will act as a junior partner and critic, keeping track of details and assisting in the documentation, verification, debugging, and modification of a program while the programmer does the really *hard* parts of design and implementation. In order to cooperate with a programmer, the PA must be able to *understand* what is going on." [Waters82] (Italics by Waters).

The programmer's apprentice is essentially a reactive entity. It does not anticipate what it will be asked or expected to do next and is always given a general outline on which it elaborates. Typical commands to a PA are (from [Waters82]).

```
Define a program SQRT with a parameter NUM.
Implement the test as an equality within epsilon.
Share the / in the test and the second / in the approximation.
```

Automatic programming systems always seem to boil down to simply being *high level programming languages*.

"In short, automatic programming always has been a euphemism for programming with a higher-level language than was then available to the programmer. Research in automatic programming is simply research in the implementation of higher-level programming languages." [**Parnas85**](p. 438)

The programmer's apprentice is difficult to identify as a high level programming language because the programmer's attention is focussed on the object code. The programmer's program is not written in "Programmer's Apprentice", but rather in LISP with the help of a clever macro expanding programmer's apprentice. A similar system could be constructed one level down for assembly language and Pascal[**BSI82**] . The programmer would work on an assembly language program but could type in small portions of Pascal code which would immediately be translated into assembly language and inserted into the program. At that point, the Pascal code (instructions to programmer's apprentice) would be forgotten and the programmer would manipulate only the resultant assembly language.

The Programmer's Assistant[Teitelman72][Teitelman84] is an excellent example of an added-on outer environment. All communication from the user to the inner environment passes through the programmer's assistant. The programmer's assistant provides a variety of aids. These are, the maintenance and access of a command history list, the capacity to "undo" commands, and DWIM automatic spelling correction[Teitelman72][Teitelman72]. The way that the programmer's assistant presents itself to the user (usually as completely transparent) is very like the way that a user predictor presents itself.

User predictors are similar to the programmer's assistant. However, user predictors perform a completely new and independent function; there is nothing in the armory of the programmer's assistant that predicts the user. Nevertheless, automatic prediction is a function that would considerably enhance the already powerful facility that the programmer's apprentice provides. *Thus, prediction systems can be slotted in as another string to the bow of the programmer's assistant whilst being welcomed as a new innovation.*

It is interesting to take a cursory look at some recent commercial products that exploit the programmer's helper idea. One such product is *Turbo Lightning* from Borland International[**Borland86**]. Turbo Lightning is a background process that intercepts the user's input stream and looks up every word typed (it is presumed that a word is defined to be a sequence of letters) in a dictionary. If it sees a word that it doesn't know, it interrupts the user with a menu showing a list of candidate corrections. The user can then choose a correction or define a new word.

> "So how does it work? Let's say the word you meant to type was 'RIGHT' but you accidentally typed 'RIHGT,' which is wrong. What happens then? You immediately hear a 'beep,' so you know there was a boo-boo. You instantly see a window, that doesn't list "RIHGT" but it does list 'RIGHT' and its sound-alike words." [Borland86]

This product is significant because it indicates that the demand for environment assistants is strong enough to make them commercially viable. Many similar products have appeared since Turbo Lightning.

6.5.4　Research into Command Languages

Much work has been done on command language design (e.g. [Palme79] ("User Commands"), [Hardy82] and [Nievergelt82]). Command languages are important because every user has to face them to accomplish anything. Researchers in this area are concerned with command abbreviation, menus versus command lines, the trade-off between power and ease of use and flat versus hierarchical command structures. These issues are relevant to user predictors only as far as they describe much of the domain of user/computer discourse that user predictors will be charged with predicting.

Of particular interest are command recall facilities, which allow the user to retrieve and re-enter commands previously given.

6.5.5　Modelling the User at the Keystroke Level

A lot of work has been done in the area of low level modelling, (e.g. [Card80](p. 401) and [Roberts83]). Workers in this area construct models of user behaviour at the level of individual physical actions such as moving a mouse or pressing a key, and attach timing information to these actions. The result is a model that predicts the time it will take for users to perform particular sequences of actions. This information can then be used in the design of text editors and other interactive systems. The most typical of these papers is [Card80] in which an elaborate keystroke level model is built that functions as a fairly good timing-predictive model. Other related work in relation to text editors is [Roberts83].

Most of this work is not relevant to user predictors as it deals with short sequences of keystrokes and the timing of low level user actions. Although some writers delve into the correlations between various letters, most describe models that predict only the time that an action will take, not the action itself.

6.5.6 Modelling the User at the Command Level

As with the research described in the previous section, research in this area is concerned with *performance times* (e.g. [Booth81] (Esp. "Modelling the Task"), [Zunde81]). However, the problem has been raised a level. Here the concern is the performance of users performing higher level tasks for which they may choose their own commands. The result is a more complex analysis, parts of which are relevant to user predictors. In [Booth81] for example, the authors describe a modelling method by which a grammar is constructed to imitate the user's use of a software tool.

Some work is not concerned with performance times but rather with user customization. In **[Rich83]**(p. 203), the author asserts that computer systems should behave in different ways for different users. Two approaches for achieving this are given. First, the user can do the customizing. This can be complicated and does not cater for new users. Second, the customization can be performed automatically. The rest of the paper continues on this theme indicating how clues from the user behaviour can be used to make using the computer easier. This is relevant to user predictors as such a system uses information gleaned from the user to make using the computer easier.

In [Fischer84], the authors describe a help facility that models the user and occasionally jumps out at the user and describes what the user is doing wrong and how it can be fixed.

6.5.7 Summary of Interactive Environment Work

An increasing emphasis is being placed on increasing the productivity of the users of computer systems. This has resulted in sophisticated interactive environments that can be modelled as two-level systems composed of an inner environment containing the objects that the user is manipulating, and an outer environment consisting of a collection of tools whose function is to help the user use the inner environment. Prediction systems can be categorized as tools residing in outer environments. A lot of work has been done on interactive programming environments. Much of this is not relevant to user predictors. The studies on outer environment tools have been helpful. User prediction systems can be regarded as part of the tool kit of the programmer's assistant.

6.6   Multi-Character Predictions

As we have seen, user prediction systems operate in a manner similar to that of data compression systems. As a consequence, all of the models and methods for predictions discussed in earlier parts of this thesis are applicable. The major difference between the two systems is the kind of predictions they produce. Whereas the existence of arithmetic coding allows data compression models to predict a single character, user predictors must predict whole strings of characters to be at all useful.

The only useful user predictions are those that would normally take more than one keystroke to type. The cost of the user reading the prediction to determine if it is correct sets the threshold of usefulness even higher. Predictions of less than five characters are probably useless. This means that in order for useful user-prediction to take place, the user input must be highly redundant. It must be possible in some circumstances to predict whole strings of characters with confidence. Predictions of low entropy (e.g. because there are only four symbols) but uniform probability are of little use.

String-level redundancy can be measured by identifying all repeated substrings in a section of user input. Experiment 22 in Appendix C performed such a measurement. The experiment showed that in the sample of 32948 consecutive instances taken from a user input stream, at least a third of the instances lay within a repeated substring of length ten or greater. This result indicates a high degree of string-level predictability. Experiments using the mechanisms of earlier chapters have not been performed.

Multi-character predictions can be obtained from a model that only predicts a single character by invoking the model recursively. In general, a multi-instance prediction of length k consists of a n-way tree that is solid to depth k, with a probability on each arc. Such a tree can be generated using a single character predictor by feeding characters into the predictor and pulling out a hypothetical prediction. If the current context was beg, the probabilities on the arcs leading from the node begi could be obtained by temporarily feeding the character i into the model and then examining the prediction yielded.

In practice, a user predictor would never actually build a prediction tree. Even if such a tree could be efficiently constructed, it could not be presented to the user, because the cost of reading the tree would exceed the cost of typing the characters the tree predicts. At each point of time, only a very few predictions can be used. These predictions can be found by following branches of high probability, constructing them on the fly. The difficulty then becomes that of deciding which predictions from the prediction tree should be chosen for presentation to the user. The trade-off here is between length and probability. Should the user be presented with short, reliable predictions or with lengthy long-shots?

The problem can be expressed formally as the need for a goodness function $g(l,p)$ that can be applied to each node in the tree, where l is the depth of the node (length of the prediction in characters) and p is the probability of the node arising.[120] Once this function is specified, the problem becomes that of simple tree search. About all we know about g is that it must be monotonically increasing with l and with p.

A good first approximation is $g(l,p) = p(l-1)$ which is the number of keystrokes saved. If the cost of examining the prediction is taken into account, the metric can be refined to $g(l,p) = p(l-1) - (i+sl)$ where i is the cost in the user's time of being interrupted to be given the prediction and s is the cost per character of reading the prediction.

Once a goodness function has been found, an algorithm is required for finding the tree node with the greatest goodness. A branch and bound algorithm[**Winston77**] seems appropriate but can only be used if a limit is placed on the length of predictions. The best way to do this would be to modify g so that its value increases and then decreases with increasing l.

If more than one prediction is required (say the k best prediction strings), the branch and bound algorithm could be modified to maintain a list of the best k predictions with the cutoff point being set at the worst of the k predictions.

The amount that should be predicted is inextricably linked with the structure of the commands the user is giving. From a psychological standpoint, the best thing that the prediction system could do is to predict one or more chunks of user action. For example, a chunk could be a complete command. Luckily, the boundaries of such commands will tend to correspond with the points where the user will make a choice about what is to be done next. These branching points are, in turn, reflected by an uncertainty in what to predict next. Thus, command boundaries can be detected from high entropy branchings in the prediction tree.

---

[120] The method of assigning probabilities here is the same as that used by Rissanen and Langdon[Rissanen81] in their definition of a source.

One source of information available in a user prediction system but not in a data compression system is the timing of the arrival of the characters. There are many reasons to think that much of the structure of an input could be determined from the time intervals between the keystrokes.

- The existence of muscle memory means that commonly typed sequences are likely to be typed in a burst.

- When users type a command terminator, they have to wait until the computer responds. The time taken by the computer will delay entry of the next keystroke.

- When users reach the end of a coherent conceptual unit, they pause.

Some of these hypotheses have been tested by writing a program to tokenize an input stream based only on the time interval between keystrokes. The result is presented in Experiment 23 in Appendix C. Although no statistical tests have been performed upon the parsed output, it appears by inspection that the program has determined much of the structure of the input stream.

6.7   The Prediction Interface

Given k prediction strings, how can they best be presented to the user? We divide the interaction with the user into two parts: the part that presents the prediction to the user and the part that reacts to the user's response to the prediction. Each of these can be rated on an aggressiveness scale.

6.7.1   Presenting Predictions

The presentation of predictions involves somehow making the user aware that a prediction exists so that action can be taken. Here are some approaches rated from least aggressive to most aggressive.

- The least aggressive thing that a prediction system can do is absolutely nothing.

- The next least aggressive thing that a prediction system can do is to present predictions to the user only when the user requests them. There could be a special key on the keyboard for this purpose. After a while, the user would gain an intuitive sense for the situations in which a prediction system is likely to be correct.

- The next presentation option is for the prediction system automatically to present the user with a prediction whenever the system has a prediction with a high goodness.

- A fourth method of presentation is to continually present the prediction system's best prediction. Window systems would be well suited for this.

- The most aggressive system would continually present a list of the best predictions.

6.7.2 Confirming Predictions

When a prediction is presented to the user, the user must respond to it in some way. One response is to do nothing. Here are some of the ways in which the user could confirm predictions. They are rated from least aggressive to most aggressive.

- The least aggressive method for confirming predictions is to provide a key for confirming a prediction. The user could press the key if he wanted the prediction executed, or could ignore it and continue.

- A second and more aggressive option is to set up a key that indicates that the user does *not* want a prediction executed. If, after a prediction is presented, the user does not respond in a prescribed time, the system executes the command.

- Finally, there is the extremely aggressive option of simply executing the prediction without confirmation. Extreme as it sounds, this option could be very useful in situations when the goodness value of a prediction is extremely high.

Our experience with prediction systems indicates that introverted prediction systems are the best as they tend to fit in with the concept that the user is in the driving seat.

Perhaps the best choice amongst these is all of them. The prediction system could employ a mixture of these and choose any one method depending on how sure it is of its prediction.

6.8 Tuning Models

If Markov models are to be used to predict the user, they must be tuned. In this section we note the parameters that we expect would perform well.

The two properties of user data that should be kept in mind are first that commands that have just been issued are often repeated a short time afterwards, and second that commands are often quite long. We work on the principle that in a user prediction system, a high entropy prediction is a completely useless one. Branches that contain high entropy predictions may as well be pruned.

The maximum depth of the tree should be set much higher than for data compression. In data compression the optimal depth is about four; deeper trees do little to reduce the entropy while making the tree too specific. In user prediction, this specificity is likely to be useful because users often repeat commands soon after they are given. In a user prediction system, the cost of being wrong (and being wrong is a much sharper concept in these systems) is much higher than for data compression.

Tree growth should be set to be very high. Because the user's most recent input is likely to be of greatest relevance, the tree growth parameters should be adjusted to grow each branch to the depth limit upon the arrival of each character. Windowed local adaptivity should probably be turned off. The estimation parameters should be biased strongly against symbols that have not occurred (i.e. a low value of λ).

One of the pleasing aspects of the tree adjustment techniques presented in Chapter 4 is that they are incremental. Whereas most previous algorithms destroy the Markov tree and start anew whenever memory runs out, SAKDC can operate smoothly forever. Use of such incremental tree adjustment means that there are no discontinuities in the performance of an SAKDC user predictor.

Finally, we note that the user input stream is likely to be multimodal. Different software tools have different input grammars and are likely to have different characteristics. As the user switches from one software tool to another, so should the compressor switch from one model to another. This could be done automatically without the need for a tight link between the software tools and the prediction system.

6.9 Human Factors

Interactive systems are extremely prone to instability with respect to user perception. Such little things as the exact length of response time or the phrasing of error messages can greatly affect the attitude that the user takes towards the system. In introducing a system as unusual as a user predictor, it is prudent to take at least a cursory look at the psychological factors involved.

Perhaps the single biggest difference that a user predicting system could make to an interactive environment is to make the user no longer feel in control. Traditionally, human/computer interfaces are set up as master/slave relationships. The computer waits until the human types in a command and then executes it. It displays some messages to the user and then waits for another command. This is a user-driven dialogue. In contrast, some user prediction systems might actually interrupt the user with suggestions or possible commands to be executed. Users who are used to the computer "not speaking until spoken to" may find this behaviour disconcerting.

Humans do not like to be predicted because knowledge of their own predictability compromises their self-image as agents of free will [Dennett84]. People hold the art of prediction in high regard and can feel threatened if successfully predicted. An experiment has shown that resentment is generated by people whose predictive powers (of human performance) have been upstaged by a machine[Dawes71].

A similar problem of some user prediction systems is the way in which they "suggest" a course of action to the user. For example, the user predictor might predict that the user is about to edit a program and suggest this action when, for the first time in twenty compilation iterations, the user actually wants to run the program. This sort of prediction could be interpreted by the user as bossiness, making the user feel pressured by the system to perform a particular course of action. Witten and Cleary[Witten86] touch on a similar point in their paper on general applications of predictive models.

> "The prediction methods suggest continuations that have occurred frequently in the previous text (or the priming text). It is dangerous to use the predictions as *suggestions* of what to type; for the result will lack variety, vigor, and verve. Instead, it is essential to conceive what is to be entered *first*, and use the predictions to facilitate its entry. Otherwise, these techniques will encourage stultification, unimaginative prose, and we will want to disown them." [Witten86](emphasis by Witten)

A more positive way of viewing the warning is to realize that the predictor is probably a very good indication of what *not* to write. If good writing is the aim, the user might do well to aim to *maximize* the number of keystrokes! In contrast, many dialogues with computers are structured by the computer software and creativity is not a factor. In this case, suggestions are likely to be welcome.

It is interesting to note that here the computer is pressuring the human to conform in the present to the image that the human has projected in the past. This sort of pressure could have a powerful effect on users as it is the way that much social interaction occurs. Individuals, in interacting with a group of other people, develop a social personality that the group feels happy with. Any deviation from this personality, often even if it is for the better, is met by strong resistance by the group [Schachter51]. Self correcting mechanisms and reinforcers of social images such as nicknames are used to create stability.

To summarize, a prediction system, in suggesting that a particular course of action is to follow, may be interpreted by the user to be implying much more. In the extreme case of someone who has never used a computer before, such a prediction could be interpreted as a demand. It may be possible to counter these effects by careful wording, explanation and by manipulating the user's perception of the capabilities of the prediction system. It is reassuring to see that automatic helpers are becoming more common in interactive environments, at least for environments used by experts, as this indicates that in practice these effects are at least not prohibitive.

6.10 Work by Witten, Cleary and Darragh

Most of this chapter was written in late 1986. In mid 1988, the author of this thesis became aware of work performed by Witten, Cleary and Darragh in 1982–1983 in the area of user prediction. Their work, which duplicates many of the ideas in this chapter, is described in two papers, the first by Witten[**Witten82**] and the second by Witten, Cleary and Darragh[**Witten83**].

In the first paper, Witten[Witten82] described a line-based user-prediction system implemented under Unix. As the user types each Unix command, the prediction system displays its prediction of what the user is about to type next. The prediction is presented in inverse video at the end of the user's line. If the user does not agree with the prediction, the user continues typing as if no prediction had been made. If the user agrees with the prediction, one or more function keys can be pressed to transmit all or part of the prediction. The system does not attempt to predict past a newline character.

The predictor operates in one of three modes: character, word or string, depending on the level at which it tokenizes (forms symbols), and it uses a fixed-order Markov model of the symbols. If a context contains a single symbol, that symbol is predicted, otherwise a choice is made. One strategy investigated for this case was to choose the symbol with the most recently added instance, unless another symbol had a significantly higher frequency count.

Witten experimented with the granularity of tokens and the order of model, and found that an order three, character-level model performed best. Performance decreased slowly from order three to order twenty. This is consistent with the findings of Experiment 5 of this thesis.[121] Witten experimented with a symbol-level credibility threshold in which a symbol's frequency was treated as zero if it was below a certain threshold. This technique reduced the error rate but also reduced the number of predictions.

In general, Witten's user predictor worked well, successfully predicting about one quarter of the user's input. However, Witten seemed to think that 25% was a small proportion of the user input to predict.

> "This may still be useful for poor typists, with around 25% of characters being predicted correctly." [Witten82]

Witten stressed that the system is not likely to be useful to experienced users and fast typists. He also stated that the system could not be used with programs that do not echo their input.

In contrast, we feel that such systems will be of great use to experienced users. For a start, experienced users are used to working with a multitude of tools and could assimilate a user prediction system into their environment with little fuss. Beginners often have difficulty understanding that a computer system contains many different software components and find it even harder to distinguish between the components.[122]

---

[121] However, these results contrast with the author of this thesis's own estimates of the best depth (Section 6.8).

[122] Each year, at the Department of Computer Science at the University of Adelaide, hundreds of first year students type command interpreter commands (such as commands to compile) into the screen editor and then raise their hand for help when nothing happens.

Second, the system need not be used for *all* input. It need be used only when the input to the system becomes particularly monotonous. This often happens when a "one-off" task (such as moving one hundred files from one machine to another) must be performed, and it is not worth writing programs to perform the task. A prediction system can fill the gap between typing in commands and writing programs. We might expect the cost structure to look something like **Figure 101.**

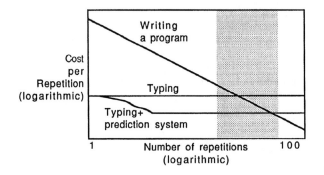

User prediction systems could be used to fill the efficiency gap between issuing commands directly (e.g. giving Unix commands) and constructing programs to issue them (e.g. writing and executing a Unix shell script containing a loop). If a command is to repeated only a few times, it is best typed directly. If a command is to be executed thousands of times, the overhead of writing a program becomes negligible compared to the cost of typing the commands. Prediction systems could fill the gap between these extremes.

Figure 101: Cost of typing and programming vs repetition.

Third, it is quite possible to use prediction systems in non line-oriented interfaces. All that is required is that a printable representation be found for non-printable characters so that predictions can be presented on the screen. In fact, it is likely that screen-based interfaces will account for most of the use of a prediction system. It is common, for example, in screen-based editors to wish to perform a complicated operation repeatedly on a number of lines.[123]

At the end of his paper, Witten hints at the multimodal nature of user input:

---

[123] In early 1986, the author of this thesis constructed a prediction system using what was essentially a high-order, variable-order Markov model. The system was slow (about 10 characters per Vax750 CPU second) but predicted editor command sequences as effectively as anything else.

"Further research is needed to assess the benefit of continuing the thread of *predict*'s context right through the interactive dialogue, irrespective of the subsystems that the user enters. It may be preferable to save *predict*'s state on entry to subsystems and preserve a context from one invocation of a subsystem to the next." [Witten82](last paragraph)

This quote touches on an important aspect of multimodal systems. Although the algorithm of Chapter 5 is capable of determining the sources just from looking at the data stream, in practice there are times when source transitions may actually be known.

Witten's first paper presented the idea of user prediction and described a very specified Unix line-based implementation of a predictor. In the second paper, Witten, Cleary and Darragh[Witten83] presented a more sophisticated and finely grained predictor that operates using a menu. Whereas the 1982 predictor presented a single prediction, the later predictor, called "the reactive keyboard", presents a number of prediction strings organized into a menu. Whereas in the earlier predictor, predictions were confirmed in full or in part by pressing a function key, in the reactive keyboard a prediction is confirmed by clicking part way into the prediction with a pointing device. That part to the left of the click position is entered. This means that for a menu with ten-entries of ten characters each, there are 100 possible selections. Menu items are listed in probability order. These innovations partly solve the problems presented in Section 6.6 of having to decide how many predictions to make and how much to predict.

Despite the earlier results which indicated that low orders were the most effective, the reactive keyboard uses a PPM model of *order ten*.

Witten, Cleary and Darragh experimented with different sized menus and produced some graphs showing the costs of various sized menus. Once the menu size reaches about 20 entries, there is little to gain by increasing it. Finally, they discussed the other possibilities presented by the technique. These include spelling checking and other constraints and the ability to present non-text symbols.

6.11 Summary

This chapter has introduced user-prediction as a new application of data compression techniques. The modern paradigm of data compression has been modified for use in predicting users. Some examples were given of how a user prediction system could be used. A review of the field of interactive programming environments revealed that environments can be divided into two parts, an inner environment and an outer environment. User predictors are outer-environment entities suitable for incorporation in Teitelman's programmer's assistant system.

Arithmetic codes allow data predictors to predict a single character. In contrast, user prediction systems must predict more than one character. By recursive invocation, a single-character predictor can generate a multi-character prediction tree from which the best predictions can be selected using a branch and bound search algorithm and a goodness function. The best predictions can be presented to the user in a variety of ways.

User input differs from other data in significant ways, and the requirements on a user predictor are stringent. First, the system must operate in real-time, without lengthy delays. Second, the predictor must adapt to the user's behaviour. Third, user input is likely to be multimodal. The incremental, adaptive algorithms described in early chapters are ideally suited to meet these requirements.

Human factors are important in a user prediction system; the user must be made to feel in control.

Finally, the work of Witten, Cleary and Darragh has been acknowledged. Although their work is slightly pessimistic about the applicability of user prediction, we have given good reasons for optimism: that experienced users will be comfortable with user predictors, that prediction systems are applicable to screen based programs, and that user predictors fill in a productivity gap between typing commands manually and writing a program to execute them.

Chapter 7
Conclusions

The goal of this research was to investigate the use of adaptivity in data compression. This goal has been achieved by identifying different kinds of adaptivity, implementing them and evaluating their performance. This chapter presents the highlights of this thesis. It does not summarize it. Summaries appear at the end of each chapter.

7.1 Primary Contributions

Primary contributions of this thesis are listed below.

- A comprehensive literature review of the field of data compression has been given (Chapter 1).

- The concept of adaptivity in a data compression algorithm has been refined by focussing on the emphasis that a particular technique places on different parts of the history, and by constraining the estimation function. Four different classes of adaptivity have been identified, and mechanisms for each form of adaptivity given (Chapter 3).

- A general Markov algorithm (the SAKDC algorithm) has been presented that integrates many diverse features of contemporary Markov algorithms, and implements local and asymptotic adaptivity at the context and structure levels (Chapter 4).

- Experiments on the SAKDC algorithm have been conducted. The results will be useful to practitioners who are tuning a production Markov compressor (Section 4.17).

- The experiments show that for small memory (less than 5000 nodes), structural adaptivity (in which nodes are moved within the Markov tree) can improve compression by up to 20% absolute (Section 4.17.15).

• The techniques of suboptimality and incarnation numbers have been introduced as a solution to the problem of maintaining short-cut pointers in a tree whose structure is changing (Section 4.10).

• The class of multimodal sources has been identified. The class is interesting because it is likely to appear on communication lines. A multimodal data compression algorithm (MMDC) has been described that outperforms ordinary models for these sources, by creating and maintaining a model for each detected source mode (Chapter 5).

• An application of data compression models to user interfaces has been proposed. A model of the user's input stream is developed and used to predict future input. The incremental, adaptive techniques described in this thesis are ideally suited for this application (Chapter 6).

7.2 Secondary Contributions

The secondary contributions of this thesis are listed below.

• The distinction between "symbols" and "instances" has been introduced (Section 1.2).

• The estimation techniques used in previous Markov algorithms have been generalized and classified into two groups, linear and non-linear, with a λ parameter (Section 1.10.3).

• A generalized form of floating point, stochastically incremented counters has been introduced (Section 1.11.6).

• The relationship between backwards and forwards trees has been analysed and the advantages of each enumerated (Section 1.12.2).

• Problems in the literature in reporting compression performance results have been identified and a nomenclature proposed (Section 1.15).

• A method of retarding Markov tree growth using thresholding has been given, and upper and lower bounds on the growth rate derived (Section 2.7).

- The distinction between context adaptivity and structural adaptivity has been made (Section 3.7).

- The two colour LRU management problem has been identified and solved in logarithmic time (Section 4.7.1).

- A method for organizing heaps using dynamic memory allocation has been given (Section 4.7.2).

- A mechanism for structural and contextual windowed local adaptivity has been given (Section 3.9 and Section 4.13).

- The concept of source zones has been developed as a possible explanation to the behaviour of the MMDC algorithm (Section 5.7.5).

- Possible security problems arising from sophisticated data compression techniques (and in particular Multimodal data compression) have been identified (Section 5.8.6).

- A study of the word "adapt" has been made (Appendix B).

- An experiment was performed that indicated that user input is highly redundant (Section C.2).

- An experiment was performed that indicated that there is a strong relationship between the pauses between keystrokes and the characters being typed. An input stream was tokenized solely on the basis of timing information (Section C.3).

- Ideas for further research are given in Appendix D.

7.3 About Communication

Insofar as data compression is a special case of data representation, this thesis has been about *communication*. Whereas much communication can be viewed as a simple process in which information is transmitted using a fixed, well-defined alphabet having fixed, well-defined meanings, the modern paradigm of data compression presents a much less stable picture in which the semantics of the channel alphabet change from moment to moment. The only way to make sense of such communication is to refer to the predictions that underpin it. Here we consider predictions as defining the set of events and their semantics as well as their probabilities.

These predictions, which are so changeable, map channel symbols to semantics, (which in the case of computer communication consist simply of the set of source symbols). Without predictions, communication involving representation[124] cannot take place. Communication cannot occur without the two parties somehow setting up a system of predictions to determine the semantics of the communication.

In ordinary communication using a simple-alphabet, both sides agree to attach simple, fixed semantics (fixed prediction) to the channel alphabet. In a data compression system that uses the modern-paradigm, both sides agree upon a method for generating predictions from what has passed before. In user prediction systems the computer explicitly informs the user of the predictions (through a special backwards channel).[125]

In summary, the principles underlying this thesis highlight the adaptive nature of most communication and lead to a view of communication that focuses on the generation of predictions rather than the generation of instances.

7.4 Towards a More General Theory

Much of the work in data compression has been pre-scientific. This thesis is no exception. Much of what has been achieved has been driven by pragmatic concerns backed by intuitive theory. There are two lights on the horizon which may be worth following in order to arrive at a more systematic approach to the field.

The first is the tantalizing correspondence between multimodal data compression and Markov models. Each works with "contexts", one at the instance level and one at the ten-thousand instance level. The Markov model concept of a context is that of a short context string. The multimodal concept of a context is that of a mode of behaviour over a long period. Ideally these two concepts could be fused into a single compressor that incorporates the concept of context at many levels, possibly recursively.

Second, many of the estimation and adaptivity techniques discussed in this thesis have been on the fringe of signal theory. The field of text data compression may well be at the point where it can profitably

---

[124] All communication at the physical level is rooted in the semantics of reality common to all objects; no predictions are required. If a tree branch (from a Markov tree of course) falls on me and kills me, it communicates with me and the semantics are clear.

[125] For this to take place, the computer and the user had to "agree" upon a protocol for interpreting such information.

employ concepts from this field. For example, we might think of the mode changes of a multimodal source as having a frequency spectrum. If a generalized concept of context were to arise, it is possible that the concept of frequency spectrum of change could be applied more generally to a source, resulting in a better understanding of the systems being studied.

7.5 Thesis Perspective

Much of the author's candidature has been spent exploring and catching up with the fields of interactive programming environments and data compression. The thesis has grown out of this process.

The following table indicates the path that the author has taken to define this thesis. The first two columns list an idea that the author had and the date at which the author had the idea. The third column lists the paper that introduced the idea (or a similar idea) and the date that the author discovered the paper.

Date	Idea	Previously published by
Feb–85	Prediction of user input	Witten82 (Jun–88)
Oct–86	Predict/code paradigm	Rissanen81 (Sep–87)
Oct–86	Markov-tree modelling	Cleary84 (Sep–87)
Oct–86	Arithmetic coding	Several (1963–1987) (Oct–86)
Nov–87	Adaptive mechanisms	Original
Oct–88	Multimodal algorithm	Original

While inefficient, rediscovery has resulted in different perspectives. For example, the author's first Markov tree algorithm (DHPC) employed an estimation technique inferior to that previously published but used a more sophisticated tree management technique. Thus, this thesis has approached old ideas from a different direction as well as making some original contributions.

7.6 Summary

This thesis has made a number of contributions to the field of data compression. The concept of adaptivity has been refined by defining it in terms of constraints on the estimation function ξ and the history weighting function w. Four classes of adaptivity were identified, as were mechanisms for implementing each class. Some of these mechanisms were incorporated into a general Markov algorithm. Experiments on the algorithm showed that adaptivity can improve compression considerably. The class of multimodal sources was identified and an algorithm described that performs better on such sources than previous algorithms. Finally, a proposal was given for the application of data compression techniques to user interfaces.

Appendix A

Estimation Formula Calculation

Although there is no theoretical basis for deciding between different estimation formulae (Section 1.10.3), it is possible to derive the best formula for a given meta-distribution (distribution of distributions). The author of this thesis has derived the best estimation function for the uniformly meta-distributed binary memoryless source. The solution was linear estimation with $\lambda = 2$.

Jones (C.B.) has given a more general derivation (for more than two symbols) which is reproduced here exactly.[126] The solution, which is consistent with the author of this thesis's binary analysis, is linear estimation with $\lambda = n$.

Consider first a binary source with probability p of one and q of zero. The probability of a sequence of length n containing r ones is

$$b\left(r, n, p\right) = \binom{n}{r} p^r q^{n-r}$$

where the binomial coefficient $\binom{n}{r} = n! / (n-r)! r!$ Suppose that all possible binary sources are equally likely, i.e. the (unknown) probability p was chosen randomly in the range $(0, 1)$. The probability of a sequence of length n containing r ones is then

$$b\left(r, n, p\right) = \binom{n}{r} \int_0^1 p^r \left(1 - p\right)^{n-r} dp$$

---

[126] Private communication (letter) 7 September 1987. Permission to reproduce the derivation here was granted by telephone on 20 April 1989. Note: Because this proof is reproduced exactly, the notation is different from that of the rest of the thesis.

The integral has the form of the Beta function[**Abramowitz72**] and

$$b(r,n,p) = \binom{n}{r} \beta(r+1, n-r+1)$$
$$= \binom{n}{r} \Gamma(r+1) \Gamma(n-r+1) / \Gamma(n+2)$$

Substituting for the Gamma function $\Gamma(n+1) = n!$

$$b(r,n,p) = \frac{\binom{n}{r} r!(n-r)!}{(n+1)!} = \frac{1}{n+1}$$

Thus it is equally probable that a sequence of length n contains $0, 1, 2, \ldots, n$ ones and the probability of a particular sequence containing r ones is

$$\frac{1}{(n+1)\binom{n}{r}} = \frac{r!(n-r)!}{(n+1)!}$$

This is presented as equation (4) in [**Lawrence77**]. When a particular sequence of length n containing r ones has been received, if the next symbol is one there will be a sequence of length $n+1$ containing $r+1$ ones, so the conditional probability of one is

$$\frac{(n+1)\binom{n}{r}}{(n+2)\binom{n+1}{r+1}} = \frac{(n+1)!(r+1)!(n-r)!}{(n+2)!r!(n-r)!} = \frac{r+1}{n+2}$$

This is the probability estimation function for a binary source. The generalization to a source with k different symbols proceeds as follows. After n symbols, let the number of occurrences (≥ 0) of each symbol be

$$r_1 + r_2 + \ldots + r_k = n$$

The number of distributions satisfying the above equation (see [Feller57]) is

$$\binom{n+k-1}{n}$$

and the number of the ways n symbols can be partitioned according to a particular distribution is

$$\frac{n!}{r_1! r_2! \ldots r_k!}$$

By analogy with the binary source, we assume that each distribution is equally probable. The probability of a particular sequence of n symbols (with a particular distribution) is therefore

$$\frac{r_1! r_2! \ldots r_k!}{\binom{n+k-1}{n} n!} = \frac{(k-1)! r_1! r_2! \ldots r_k!}{(n+k-1)!}$$

The conditional probability that the $n + 1$th symbol will be symbol i is then

$$\frac{r_1!r_2!\ldots(r_i+1)!\ldots r_k!}{r_1!r_2!\ldots r_i!\ldots r_k!} \cdot \frac{(n+k-1)!}{(n+k)!} = \frac{r_i+1}{n+k}$$

This then is the probability estimation function for a source with k different symbols.

Appendix B

The Word "Adapt" And Its Forms

During the writing of this thesis, it became clear that the word "adapt" is a problem word. For a period in the preparation of Chapter 3 chaos reigned with different forms of the word "adapt" being used interchangeably. Eventually, the author decided to address the issue explicitly.

The word "adapt" has many forms, many of which are interchangeable. **Table 50** contains a list of all the forms of "adapt" that the author has found. All the words appear in the Oxford English Dictionary[**Murray33**] (herewith "the OED") except for "adaptably" and "adaptivity", which are marked with a dagger (†). The meanings are paraphrases of the OED definition but the examples are by the author.

To sort out these alternatives, the words were arranged into groups that compete for grammatical slots (**Table 51**).

A few notes on the table of competing words are in order. The verb forms have few variations. The form "adaptate" is rarely used. The words "adapter" and "adaptor" are merely different spellings of the same word with "adaptor" listed as rarely used. The OED's 1972 supplement lists the form "adaptor" as now commonly used to describe electrical fittings. Of the group "adaptation", "adaption" and "adaptment", the commentaries *Fowler's Modern English Usage* [**Fowler65**] and *Right Words: A Guide to Modern English Usage in Australia*[**Murray-Smith87**] all indicate that "adaptation" is much more popular than "adaption". "adaptment" is rarely used.

The author's preference in each group is marked with an asterisk (*). Pluses (+) indicate alternative choices to be explained later.

With so many words competing for the same slots, one might expect semantic variations between them. However, the OED does not make any clear distinction between the different forms. In contrast, the author has found that his ear naturally distinguishes between words that indicate that an object is amenable to being adapted and words that indicate that an object is capable of adapting of its own accord. Consider the following two sentences:

"The new professor proved most adaptive."

"The pocket knife proved most adaptable."

In the first case, the object modifies itself; in the second, the object is modified by an external agent. In each case, the word "adapt" is used to indicate a degree of flexibility. By experimenting with different words in different sentences, the author arrived at the following rule:

Ad Hoc Rule: Forms beginning with **adapti-** indicate that an object is capable of modifying itself. Forms beginning with **adapta-** indicate that an object is capable of being modified.

This rule has been used to mark alternate forms in the table. Words indicating external modification have been marked with a plus (+). Here is a table of the alternate forms.

	Internal	External
Noun	adaptivity	adaptability
Adjective	adaptive	adaptable
Adverb	adaptively	adaptably

It should be emphasized that this distinction is not made in the OED, but is merely proposed by the author of this thesis. Nevertheless, with so many forms, the distinction might as well be made.

In this thesis we have attempted to use only the forms of the word "adapt" marked by * and + in **Table 51**.

Word	Kind	Meaning
adapt	verb	To alter or modify so as to fit for a new use.
		"I had to adapt the spanner to fit the nut."
adaptability	adjective	The quality of being adaptable.
		"The adaptability of the robot was extraordinary."
adaptable	adjective	Capable of being adapted.
		"Principles are adaptable to all ages."
adaptableness	adjective	=adaptability.
adaptably†	adverb	In an adaptive manner.
		"The compressor compressed adaptably."
adaptate	verb	=adapt (rare).
adaptation	noun	Noun of action. The action of adapting something.
		"We see this in a later adaptation of the play."
adaptational	adjective	Of or pertaining to adaptation.
		"Models can be divided into the adaptational and the static."
adaptative	adjective	=adaptive.
adaptativeness	noun	=adaptiveness.
adapted	adjective	Modified so as to suit new conditions.
		"The sloth is highly adapted to moving slowly."
adaptedness	noun	The quality of being adapted or suited.
		"The sloth has a greater degree of adaptedness."
adapter	noun	One who adapts other objects.
		"He was a proficient adapter of plays."
adapting	gerund	Action of rendering suitable for some purpose.
		"Adapting to foreign food is a challenge."
adaption	noun	=adaptation.
adaptitude	noun	=adapt+aptitude.
		"He lacked adaptitude."
adaptive	adjective	Characterized or given to adaptation.
		"The compressor is adaptive."
adaptively	adverb	In an adaptive manner.
		"The compressor compressed adaptively."
adaptiveness	noun	The quality of being adaptive.
		"The adaptiveness of the compressor was good."
adaptivity†	noun	The property of being able to adapt.
		"The adaptivity of the compressor was good."
adaptly	adverb	In a fit or adapted manner.
		"The miner adaptly scurried along the tunnel."
adaptment	noun	Adaptation. Fitting condition.
		"The miner's adaptment to the tunnel was eerie."
adaptness	noun	=adaptedness.
adaptor	noun	=adapter.
adaptorial	noun	=adaptive.

Table 50: The various forms of the word "adapt".

VERB past	adapted *	"We have ◊ to our circumstances."
VERB present	adapting *	"We are ◊ to our circumstances."
VERB future	adapt * adaptate	"We will ◊ to our circumstances."
NOUN for modifier	adapter * adaptor	"The algorithm is an efficient ◊."
NOUN for modifyee	adaptation * adaption adaptment	"Cutoff is evident in this ◊."
NOUN for quality of being flexible	adaptableness adaptativeness adaptiveness adaptivity * adaptitude adaptability +	"The ◊ of the algorithm is remarkable."
NOUN for quality of having been modified	adaptedness	"The ◊ of the walrus is unusual."
ADJECTIVE	adaptable + adaptational adaptative adaptorial adaptive *	"An ◊ algorithm compresses well."
ADVERB	adaptly adaptably + adaptively *	"The algorithm ◊ compressed the data."

Table 51: Forms of the word "adapt" arranged in grammatical slots.

Appendix C
User Input
Experiments

C.1 Introduction

This appendix contains a description of two experiments that were performed to investigate some characteristics of user input streams. The first experiment investigated the redundancy of user input streams and the second investigated the delays between instances in such streams. The first experiment indicated that user input streams are highly redundant. The second indicated that there is a strong relationship between the pauses between instances and the instances themselves.

C.2 Experiment 22: Redundancy of User Input

Aim: To obtain a rough measure of the predictability of typical user input.

Method: A user's terminal line was tapped and all input from the terminal was logged in a file. The delay between bytes (in milliseconds) was recorded as well as the bytes themselves. The number of bytes in the file was 32948. The user, who was a tutor in the Department of Computer Science at the University of Adelaide, knew that the line was being tapped. During the period of the tap, the user mailed mail messages, developed programs, and did many of the other things that users do.

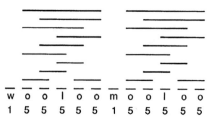

The set of all repeated substrings of a string can be represented as black bars hovering over the string. In this experiment, each instance is assigned a number, being the length of the longest repeated substring of which it is a member. Histograms of these numbers can then be plotted (elsewhere).

Figure 102: Set of all substrings in a message.

As was pointed out in Section 6.6, short predictions are of little use in user prediction. It is therefore important to find out how long the strings that can be predicted are. In this experiment every substring of length 1 or greater that was repeated in at least two different places in the input was identified (**Figure 102**). Each occurrence of each repeated substring was viewed as a black bar placed over the top of a portion of the input text, with many of the black bars overlapping. Each instance in the input was then assigned a number being the length of the longest bar containing the instance.

Results: Figure 103 is a histogram of the set of numbers associated with the instances. The horizontal axis is substring length and the vertical axis is the percentage of instances whose maximal-length repeated substring was of that length. This histogram shows that over 50% of instances fall within substrings of length ten or greater. Beyond length 30, the histogram is rather flat but still contains 25% of the input stream. Both these figures indicate a high degree of predictability.

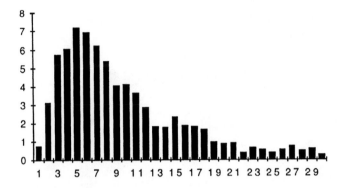

This histogram shows the percentage of instances contained in various maximal substring lengths (**Figure 102**) for a user input string of length 32948. Over fifty percent of instances fall within repeated substrings of length ten or greater.

Figure 103: Experiment 22: Percentage of lengths in raw input.

It is possible that the above result was caused by special terminal codes or sequences. To check this, the input was passed through three filters in sequence. The first filter removed escape sequences. The second filter removed instances of non printable symbols. The third filter replaced runs of instances of a single symbol by a single instance of the symbol (e.g. aaaa → a). The filtering process removed about one third of the input stream leaving 20986 instances. **Figure 104** shows the results for the filtered stream.

The filtered stream has the same characteristics as the raw stream except that most instances are contained in shorter strings. Nevertheless, the same striking statistics remain. Over 30% of instances fall within a repeated substring of length ten or greater.

The level of redundancy in the input stream becomes more obvious when the previous histograms are compared with a histogram of a random stream. **Figure 105** shows the histogram for a stream of 30000 randomly generated, uniformly distributed upper-case letters.

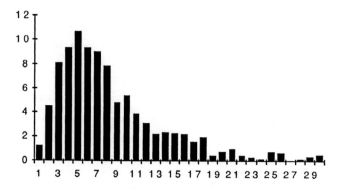

This histogram is similar to **Figure 103** but gives the results for the user input string after escape sequences, non printables and runs of identical instances had been removed. Despite the filtering, over 30% of instances fall within repeated substrings of length ten or greater.

Figure 104: Experiment 22: Percentage of lengths in filtered input.

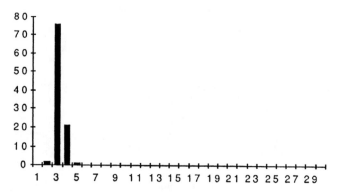

This histogram is similar to **Figure 103** but gives results for a string of 30000 randomly generated, uniformly distributed upper-case letters. The proportion of long repeated substrings is much smaller than for user input.

Figure 105: Experiment 22: Percentage of lengths in random letters.

Conclusions: Although this experiment was not performed on a large sample, the sample taken was reasonably representative. The results showed that user input is highly redundant and probably highly predictable. In this sample, at least one third of the input instances lay within a repeated substring of length ten or greater.

C.3 Experiment 23: Timing Between Keystrokes

Aim: To determine if the inter-keystroke timing intervals of a typical user input stream relate significantly to what is being typed.

Method: Two short extracts were taken from the stream captured in Experiment 22. The extracts were manually chosen but were chosen by content, not by timing. The first extract was taken from a section of the input in which the user was typing a mail message. The second extract was taken from a section of the input in which the user was typing commands in the command language (VMS DCL).

Results: In order to present the results, some method is required for representing unprintable characters. The following mapping rules were used to convert all non-printable characters into printable strings of characters.

- All printables map to themselves except '^' which maps to '^^', space which maps onto '~', and '~' which maps onto '^~'.

- Lettered control characters [1, 26] map to '^<uppercasecontrol-char>'.

- Escape maps to '^e'.

- All other characters map to '^xxx', where xxx is a decimal number that is the ASCII number of the character.

The next few pages list all the timing information for a section of input
in which the user was typing a mail message. Each line corresponds to
a single instance. Each instance appears just before the '|' on each line.
The number at the start of each line is the delay in seconds between
the instance corresponding to the line and the previous instance. The
instance appears again to the right of the '|' displaced to the right in
proportion to the delay just described.

```
00.310    ^M  |                           ^M
03.880    R   |                                                      R
00.690    i   |                                        i
00.240    g   |                   g
00.110    h   |        h
00.120    t   |        t
00.140    ~   |        ~
00.680    ^127|                                   ^127
00.350    ,   |                    ,
00.110    ~   |        ~
00.390    s   |                        s
00.120    o   |        o
00.120    ~   |        ~
00.310    n   |                  n
00.460    o   |                      o
00.160    w   |            w
00.120    ~   |        ~
01.040    w   |                                                    w
00.230    e   |                e
00.170    ~   |            ~
00.400    n   |                  n
00.340    e   |               e
00.130    e   |        e
00.150    d   |        d
00.150    ~   |        ~
00.250    t   |            t
00.230    o   |              o
00.200    ~   |           ~
00.560    p   |                         p
00.130    r   |        r
00.230    a   |              a
00.210    c   |            c
00.240    t   |            t
00.180    i   |        i
00.190    c   |           c
00.160    e   |        e
00.170    ~   |           ~
00.620    ^127|                             ^127
01.620    ,   |                                            ,
00.140    ~   |        ~
00.530    t   |                   t
00.150    o   |        o
00.110    ~   |        ~
00.970    f   |                                              f
```

```
00.150    i |          i
00.120    n |       n
00.340    d |                  d
00.140    ~ |        ~
00.310    o |                    o
00.170    u |            u
00.100    t |       t
00.130    ~ |        ~
00.640    h |                               h
00.550    o |                         o
00.130    w |          w
00.100    ~ |        ~
01.640    b |                                         b
00.330    a |                a
00.240    d |            d
00.140    ~ |        ~
02.550    w |                                         w
00.540    e |                          e
00.160    e |           e
00.890  ^127 |                                   ^127
00.180    ~ |           ~
00.480    a |                      a
00.160    r |          r
00.140    e |          e
00.300+1min. |                                    .
00.840    ~ |                              ~
03.650    S |                                       S
00.580    o |                         o
00.220    ~ |            ~
01.460    i |                                       i
00.130    f |          f
00.190    ~ |            ~
03.050    e |                                       e
00.270    v |               v
00.180    e |            e
00.130    r |          r
00.570    y |                           y
00.210    o |              o
00.330    n |                  n
00.230    e |              e
00.240    ~ |             ~
00.870    c |                                   c
00.170    o |              o
00.290    u |                 u
00.430    l |                       l
00.110    d |        d
00.120    ~ |        ~
01.370    s |                                       s
00.220    e |              e
00.320    n |                  n
00.120    d |        d
00.160    ~ |        ~
01.310    s |                                       s
00.160    o |             o
00.310    m |                  m
```

```
00.170    e |                    e
00.100    ~ |            ~
00.310    r |                           r
00.140    e |            e
00.150    t |            t
00.220    u |                 u
00.200    r |               r
00.150    n |            n
00.170    ~ |            ~
00.540    m |                                    m
00.120    a |          a
00.310    i |                       i
00.230    l |                  l
00.260    ~ |                ~
```

The information on the previous few pages indicates that the user is pausing between words and syllables. These pauses are so consistent and reliable that it is possible to write a program to *tokenize* the input based solely upon the inter-keystroke timing intervals. **Figure 106** lists an algorithm that echoes the input stream, splitting it across lines at the points where timing information indicates the end of a token. The algorithm always splits at delays greater than *break_time* and never splits at delays less than *join_time*. For delays in between these limits, the algorithm bases its decision on the average inter-instance delay since the last split.

```
join_time : constant integer ← 66;
break_time : constant integer ← 600;
renew_time : constant integer ← 300;
sum : integer ← renew_time;
count : integer ← 1;
loop
    {Delay is in milliseconds.}
    read(instance,delay);
    if (delay≥join_time) and
        ((delay≥break_time) or (delay≥2*(sum/count))) then
        write_newline;
        sum←renew_time;
        count← 1;
    else
        sum←sum+delay;
        inc count;
    end if;
    write(instance);
end loop;
```

When users type input into a computer, the delays between keystrokes are so closely related to what is being typed that it is possible to tokenize the input using only the inter-keystroke timing information to determine the breaks between tokens. The algorithm listed above accomplishes this. The algorithm has a time (*join_time*) below which it will not make a break and a time (*break_time*) above which it will always make a break. For intermediate time intervals, breaks are made if the time is greater than twice the average of the inter-keystroke intervals since the last break.

Figure 106: Experiment 23: Timing tokenizer algorithm.

Here is the output from the tokenizer for the input listed earlier in which the user was typing a mail message.

```
^M^M^M^M^M^M^M^M^M^M^M^M^M^M
R
ight~
^127,~so~now~
we~need~to~
practice~
^127
,~
to~
find~out~
how~
bad~
wee
^127~
are
.
~
So~
if~
ever
yone~
could~
send~
some~return~
mail~
to~
tell~
me~
when~
the
y~
can~
and~
can
't~
pr
actice
^127^127
se
~
(during~the~
day
)~then~
that~would~
be~
ab
so
l
utely~
ice
```

```
box
.
~ ~
Remem
ber
, ~
depending~
on~the~
timetable~
of~
matches
, ~
you
r~
first~
match~
could~be~as~ear
l~
^127
y~as~
Mon
day~
2
8th~
A
pri
ll
^127
^M
or~
it~
could~be~a~couple~
of~
wee
ks~
after~
that
.
~
~
Could~
any
ob
dy~
that~
^e[D^e[D^e[D
^e[D^e[D^e[D
^e[D^e[D^e[D^e[D
^e[C
^e[D
bo
^e[C
^e[C
^e[C
^e[C
```

```
^e[C^e[C
^e[C^e[C
doesn
't~
thin
k~they
'll~
be~
logging~
in~
every
day
^127^127^127
~day
~
tell~me~
how~
they~
can~
other
wise~
be~contacted
?
^M
```

Here is another example. In this extract, the user types a series of
DCL commands interspersed with a few commands to a scroll program.
Here is the start of the timing file.

```
00.740    ^M |                                            ^M
02.840    s  |                                                        s
00.190    c  |            c
00.130    ~  |         ~
00.340    c  |                c
00.200    a  |            a
00.260    1  |                  1
00.190    1  |              1
00.280    p  |                    p
00.360    .  |                           .
00.300    m  |                     m
00.280    o  |                   o
00.200    d  |            d
00.200    ^M |             ^M
03.350    q  |                                                        q
02.410    r  |                                                        r
00.130    e  |          e
00.250    n  |                n
00.320    ~  |                      ~
00.790    c  |                                        c
00.280    a  |              a
00.470    1  |                            1
00.180    1  |          1
00.220    p  |              p
00.300    .  |                   .
00.340    m  |                       m
00.330    o  |                      o
00.170    d  |           d
00.130    ~  |         ~
00.200    e  |            e
00.210    x  |            x
00.230    ^M |             ^M
03.210    d  |                                                 d
00.230    i  |               i
00.150    r  |        r
00.260    ^M |             ^M
02.410    ^S |                                                        ^S
13.240    ^Q |                                                        ^Q
02.150    r  |                                                        r
00.130    e  |          e
00.230    n  |                n
00.230    ~  |                ~
00.520    *  |                                  *
00.450    .  |                      .
00.200    c  |            c
00.180    o  |           o
00.300    m  |                 m
00.240    ~  |              ~
00.980    e  |                                                    e
```

```
00.250    x  |                        x
00.180   ^M  |              ^M
01.390    r  |                                                          r
00.150    e  |           e
00.170    n  |             n
00.280    ~  |                  ~
00.320    i  |                        i
00.320    n  |                      n
00.370    t  |                    t
00.250    e  |               e
00.080    r  |        r
00.450    f  |                               f
00.220    i  |              i
00.300    l  |                    l
00.100    e  |           e
01.700    .  |                                                      .
00.730    *  |                                       *
00.110    ~  |        ~
00.670    e  |                                   e
00.240    x  |                x
00.150   ^M  |              ^M
08.530    r  |                                                          r
00.100    e  |           e
00.180    n  |             n
00.340    ~  |                  ~
09.130  ^127 |                                                    ^127
00.180  ^127 |              ^127
00.200  ^127 |              ^127
00.270    s  |                 s
00.230    c  |                  c
00.190    ~  |              ~
00.490    *  |                        *
00.530    .  |                   .
00.380    d  |                      d
00.110    a  |         a
00.310    t  |                 t
00.170   ^M  |              ^M
02.820    s  |                                                     s
00.260    c  |                  c
00.260    ~  |               ~
01.370    *  |                                                     *
00.480    .  |                   .
00.220    d  |             d
00.120    a  |         a
00.250    t  |                 t
00.510   ^M  |                              ^M
07.630    q  |                                                     q
03.610    q  |                                                     q
01.860    q  |                                                     q
02.020    r  |                                                     r
00.100    e  |         e
00.180    n  |             n
00.320    ~  |                  ~
00.850    *  |                                         *
00.460    .  |                   .
```

```
00.230    d |                    d
00.100    a |         a
00.270    t |                t
00.170    ~ |             ~
01.370    e |                                        e
00.270    x |                x
00.200   ^M |          ^M
04.650    d |                                   d
00.220    i |             i
00.200    r |          r
00.240   ^M |            ^M
```

Here is what the timing tokenizer program made of the command stream.

```
^M
sc~callp.mod^M
q
ren~
callp.mod~ex^M
dir^M
^S
^Q
ren~
*.com~
ex^M
ren~interfile
.
*~
ex^M
ren~
^127^127^127sc~
*.dat^M
sc~
*.dat^M
q
q
q
ren~
*.dat~
ex^M
dir^M
^S
^Q
^S
^Q
sc~
p2.out^M
~
~
~
~
~
~
~
~
q
ren~
p2.
out~ex^M
dir^M
sc~*.mad^M
~
~
~
```

```
~
~
~
q
~
q
~
q
~
q
~
q
~
~
~
q
ren~
*.mad~
ex^M
sc~
*
.dag^M
~
~
q
ren~
*
.dag~ex
^M
dir^M
sc~
post
pr.
mod^M
^M
^M
f
~
~
q
sc~sb
disp.mod^M
~
q
ren~po
st
pr.mod~ex
^M
```

Conclusions: The timing information in user input is closely linked to the fine structure of the input. When typing in English text, the user paused at the end of words and phrases. The same effect was evident for command line input in which the user paused after commands and between words in the commands. In these experiments the relationship between the input and the timing information was so strong that the input could almost be tokenized based on the timing information alone.

Appendix D

Further Research

This appendix contains ideas for further research that arose as a result of this research.

D.1 The SAKDC Algorithm

Prediction data structure: Section 1.11 showed how important the prediction/sample data structure is in a Markov compressor. Further improvements to this structure could be made. For example, time and space performance might be improved by dynamically selecting a prediction data structure for each sample depending on the entropy of the sample; high entropy samples could use an array (e.g. Moffat's representation), medium entropy samples a binary tree, and low entropy samples a hard-coded linear search. Another alternative is to use a small hash table with linked list overflow.

Grouped prediction data structure: Langdon and Rissanen's discovery of the efficiency of approximating binary predictions by powers of two[Langdon81] suggests that the same principle could be applied to predictions of more than two symbols.[127] Predictions could be represented by $k \in \mathbf{Z}[2, \approx 10]$ disjoint sets of symbols that partition the set of symbols. The probability space ($\mathbf{R}[0, 1]$) would be divided unevenly among the sets. The probability allocated to each set would be divided evenly among its member symbols. As instances arrive, their symbols would be moved probabilistically from set to set. The allocation of probability between the sets could change dynamically as well. It might be possible to prove a tight upper bound on the inefficiency of such a technique.

---

[127] Binary predictions can be applied to sets of more than two symbols using the technique of decomposition (e.g. a binary tree of Q-Coders). Here a more direct generalization is proposed.

Estimation: Experiment 2 (Section 4.17.5) showed that dynamically modifying λ in accordance with the flatness of a sample (Section 1.10.3.3) yields a significant improvement in compression. This result, combined with the fact of the sensitivity of compression performance to the estimation formula, indicates that there is scope for further improvement in estimation formulas (and blending techniques). One avenue is the use of non-linear functions of $n - z$. Another is the use of a *local* estimate of the rate of arrival of novel instances instead of the asymptotic one currently used.

Credibility thresholds: Although Experiment 3 (Section 4.17.5) showed that credibility thresholds are detrimental to PPM blending, it also showed that credibility thresholds can improve DHPC blending by a few percent. As DHPC blending is applicable whenever execution speed is important, one research avenue is to investigate more sophisticated credibility thresholds. It might be possible to improve DHPC blending by varying the credibility threshold with the depth and possibly with the entropy of samples.

Decaying: The effect of decaying on compression is a little unclear. Experiment 8 (Section 4.17.11) indicated that there is little benefit in decaying. In contrast, other researchers have found that it improves compression [Witten87](p. 536) [Moffat88](table 4) [Abrahamson89](figure 2). Further research could resolve this discrepancy.

Hashing: Section 4.11 mentioned that if hashing was used to access nodes, it might be possible to dispense with the explicit tree structure. Instead, there would be just a collection of nodes, each corresponding to a particular context string. This would allow non-leaf nodes to be removed without restricting access to their descendants. This in turn would allow leaf nodes and non-leaf nodes to be mixed freely in an LRU list.

Longcut pointers: One of the reasons that dictionary techniques are faster than Markov techniques is that they parse and code whole strings of instances at a time. In contrast, Markov techniques parse and code each instance separately. One idea for speeding up Markov algorithms is dynamically to identify sequences of low entropy transitions in the tree and replace them by **longcut pointers**. Longcut pointers would be similar to shortcut pointers but would be labelled by a string rather than by a single symbol. The idea here is to harness the speed of dictionary techniques for the *phrase structured* ("comma dependent"[Langdon84]) parts of the source, while retaining the instance-level granularity of the Markov techniques for the non-phrase structured (comma independent) parts. Initial exploration of this idea indicates that complicated data structures might be required.

Variable-sized history buffer: SAKDC implements windowed local adaptivity by removing instances from the tree once they are K instances old (Section 4.3.4, Section 4.13). Because each arriving instance is capable of adding m nodes to the tree, the supply of nodes can run out if there are less than Km nodes. When this happens, the SAKDC algorithm switches to a different growth regime and recycles the least recently used nodes. A better approach is to work forwards from the oldest part of the history removing instances from the tree until enough nodes have been freed up to continue. This would result in a variable length history whose length would be determined by its entropy. This scheme is similar to, but not identical to, SAKDC's LRU recycling scheme.

Measuring trends: Chapter 3 showed how a weight function could be used to characterize the adaptivity of a zero-order Markov model. Each instance was multiplied by a weight and summed into a sample. Apart from the emphasis that the weight function placed on each instance, the ordering of the instances was lost. Such ordering information might turn out to be useful in measuring trends.

By treating each frequency in a sample as a signal source, predictions could be based upon the rate of change of a sample rather than on its average past behaviour. The rate of change could be calculated from estimates of the probability taken from recent and not-so-recent stretches of the history.

Random Supply: Although the heuristic "Least Recently Used" is a good one, it suffers from a very bad worst case performance. The random replacement heuristic has a poorer average performance but has no (well organized) worst case. It might be worth testing this and other heuristics in the supply system of the SAKDC algorithm.

An Optimized Compressor: The generality of the SAKDC algorithm precluded many optimizations. One research avenue is to construct a highly optimized variant of the SAKDC(Opt1) algorithm (Section 4.17.20). Among the possible optimizations are: the unrolling of loops, the use of strict local adaptivity (or decaying) to keep frequency counts down (so as to simplify coding), the use of four-byte hashing to eliminate the tree links, the use of DHPC or LAZY estimation to avoid exclusions, and the use of decomposition and binary arithmetic coding.

D.2 The MMDC Algorithm

The multimodal algorithm presented in this thesis resulted from the reasoning about adaptivity developed in Chapter 3. However, the detailed mechanics of the algorithm were arrived at through experimentation and there is much scope for improvement. Some avenues for further research are listed below.

Formal analysis of MMDC: MMDC is an algorithm looking for a theory. Theoretical analysis could be used to determine the circumstances under which the algorithm is applicable. It could also be used to determine the best parameters under a given set of conditions. One approach is to model the transitions between modes in a multimodal source using a Markov chain. Each state in the chain would correspond to a simple source. Associated with each state would be a distribution describing how long the source is likely to stay in that state. There would be a performance matrix to indicate how well the sources compress each other. Finally, a negative exponential learning curve could be assumed.

Model Management and Memory: The MMDC algorithm uses a fixed number of models, each of which receives an equal share of the available memory. This organization has two disadvantages:

- If the source has more modes than there are slots, the compressor will thrash.

- If the source has fewer modes than there are slots, some of the compressor's memory will be wasted on unused model slots.

These problems could be solved by using a variable number of models and varying the allocation of memory depending on each model's performance and activity. A danger here is that the removal of memory from a model will degrade its performance, causing a unstable feedback loop.

In the case of a source that has a single mode, a more sophisticated algorithm should allocate all its memory to a single model.

Creation threshold: The major tension in tuning MMDC is between the desire to react quickly when the source moves, and the desire to avoid the spurious creation of models. The trial period serves well here. However, further research might reveal better techniques.

Multimodal Ziv and Lempel techniques: Although the MMDC algorithm is based on the modern paradigm of data compression, it does not rely upon it. The key components of the MMDC algorithm are the maintenance of many models and the ability to switch between the models on the basis of their performance. Neither of these components require specific mechanisms. All that is required of a submodel is that it come in locally adaptive and asymptotically adaptive forms. A possible avenue of future research is the construction of a high-speed multimodal Ziv and Lempel compressor.

Contamination: In the MMDC algorithm only the local and active models are updated. When a source changes mode, the algorithm responds by switching to a different model. However, because of the negative exponential smoothing of the performance measure, there is a delay between the arrival of a new mode and the change of models. During this period, the previous model is exposed to hundreds of instances generated by the new source mode. We call this **contamination.** Further research might find ways of avoiding contamination. If the exact position of the source change could be detected (even in retrospect), the instances used to update the old model could be transferred to the new model.

Time Slicing and Model Compilation: On a sequential computer, the multimodal algorithm will take longer to execute than a monomodal algorithm. Two techniques could be used to reduce this cost.

The first technique is time slicing. Rather than obtaining predictions from every model all the time, models that are not active could take turns to generate predictions in (say) hundred byte slots. If a model performs particularly well during its time slot, it could be run in competition with the local and active model. In the general case, the models could be organized into a performance-ordered heap, with the active model at the top. Models would percolate up and down the heap according to their performance. The height of a model would determine the proportion of time that it is run, with models near the top of the heap being allocated the most time. One possible time allocation is one CPU power unit per heap level so that the active model at the top of the heap is run continually, the two models one level down are run half the time, the next four one quarter of the time, and so on. This would require time logarithmic in the number of models.

The second technique is model compilation. Most of the models in a multimodal algorithm will be inactive most of the time. It might be possible to speed optimize models that have just become inactive on the assumption that they will remain static for a while.

Anticipating mode changes: In some situations, a pattern might arise in the switching between models. By using a Markov model to model the mode transitions of the source, it might be possible to predict when transitions will occur and what the next model will be. This could be used to avoid contamination. A danger with this technique is that it might produce self-fulfilling prophecies.

Zones: Section 5.7.5 introduced the concept of tree zones to explain the behaviour of the multimodal algorithm. The presence of zoning, overlap and cross fertilization leads to the notion of a more finely grained multimodal model. One possibility is to maintain a single Markov tree that not only branches backwards (to the left) in "two dimensions" but "upwards" as well, the upwards direction being used for alternative models for the branch. Each of the vertical branchings could then be given a group number and the group numbers correlated somehow. This system would allow overlapping and cross fertilization while still allowing multiple models of some tree zones to be created.

D.3   User Interfaces

Construction: Chapter 6 of this thesis showed how the data compression techniques described in this thesis could be used to construct a user prediction system. The next stage is to build such a system and use it. Further avenues of research will then become obvious.

Appendix E
Summary of Notation

This appendix contains a summary of the mathematical notation used in this thesis. Only notation that carries meaning across a large part of the thesis is listed. Local notation is omitted. The reader is cautioned that in some parts of the thesis, symbols with global meanings are used temporarily as local symbols.

Symbol	Description
α	Extensibility threshold.
β	Credibility threshold.
ϵ	The empty string.
λ	Confidence parameter for estimation functions.
Ξ	Set of all estimation functions ξ.
ξ	Estimation function, mapping samples to predictions.
A	The set of n symbols $\{a_1 \ldots a_n\}$.
a	An arbitrary symbol.
C	Set of all sources/compression methods.
D	Set of all finite-context sources/compression methods.
d	An arbitrary source.
H	The entropy of a source.
h	The history string (h_1 is oldest element).
K	An arbitrary constant.
l	Length of a string (usually the message) in instances.
M	The set of integers $\{1 \ldots m\}$.
m	Maximum context length. Maximum depth of a Markov tree.
n	Number of symbols in the source alphabet.
P	Set of all predictions.
Q	Set of all states in a Markov chain.
$\mathbf{R}$	Set of all real numbers.
S	Set of all finite strings (of instances of symbols).
S_l	Set of all finite strings of length l.
s	An arbitrary string.
w	Adaptivity weighting function.
$\overline{w}$	Mean of a given adaptivity weighting curve.
X	The set of all samples.
x	A sample of y instances organized as n frequencies.
y	An abbreviation for $\sum_{a \in A} x(a)$.
$\mathbf{Z}$	Set of all integers.
z	Maximum number of nodes allowed in a tree.

References

[**Abrahamson89**] Abrahamson D.M., "An Adaptive Dependency Source Model For Data Compression", *Communications of the ACM*, Vol. 32, No. 1, pp. 77–83.

[**Abramowitz72**] Abramowitz M., Stegun I.A., "Handbook of Mathematical Functions", Dover Publications, Inc., 180 Varick Street, New York, N.Y. 10014, 1972.

[**Alberts83**] Alberts B., Bray D., Lewis J., Raff M., Roberts K., Watson J.D., "Molecular Biology of the Cell", Garland Publishing, Inc., 136 Madison Avenue, New York, NY 10016, 1983.

[**Altenkamp78**] Altenkamp D., Mehlhorn K., "Codes: Unequal Probabilities, Unequal Letter Costs", *Lecture Notes in Computer Science*, No. 62, pp. 15–25.

[**BSI82**] British Standards Institute, "Specification for Computer Programming Language Pascal", Publication BS6192:1982, British Standards Institute, P.O.Box 372, Milton Keynes, MK146LO, 1982.

[**Bahl74**] Bahl L.R., Kobayashi H., "Image Data Compression by Predictive Coding II: Encoding Algorithms", *IBM Journal of Research and Development*, Vol. 18, No. 2, pp. 172–179.

[**Barnsley88**] Barnsley M.F., Devaney R.L., Mandelbrot B.B., Peitgen H.O., Saupe D., Voss R.F., "The Science of Fractal Images", Springer Verlag, New York, 1988.

[**Barstow84**] Barstow D.R., Shrobe H.E., "From Interactive to Intelligent Programming Environments", *Interactive Programming Environments*, McGraw-Hill Inc., New York, 1984, pp. 558–570.

[**Bassiouni85**] Bassiouni M.A., "Data Compression in Scientific and Statistical Databases", *IEEE Transactions on Software Engineering*, Vol. 11, No. 10, pp. 1047–1058.

[Bassiouni86] Bassiouni M.A., Ok B., "Double Encoding — A Technique for Reducing Storage Requirements of Text", *Information Systems*, Vol. 11, No. 2, pp. 177–184.

[Bell88] Bell T.C., Moffat A., "A Note on the DMC Data Compression Scheme", (in press), *Computer Journal*.

[Bell89] Bell T.C., Cleary J.G., Witten I.H., "Text Compression", (in press), Prentice Hall Inc., Englewood Cliffs, NJ, 1989.

[Bentley86] Bentley J.L., Sleator D.D., Tarjan R.E., Wei V.K., "A Locally Adaptive Data Compression Scheme", *Communications of the ACM*, Vol. 29, No. 4, pp. 320–330.

[Bhat72] Bhat U.N., "Elements of Applied Stochastic Processes", Wiley, 605 Third Avenue, New York, New York 10016, 1972.

[Booth81] Booth T.L., Ammar R., Lenk R., "An Instrumentation System to Measure User Performance in Interactive Systems", *Journal of Systems and Software*, Vol. 2, No. 2, pp. 139–146.

[Borland86] Borland International, "Turbo Lightning", (Product Advertisement), *Byte*, Vol. 11, No. 2, p. 41.

[Borning87] Borning A., "Computer System Reliability and Nuclear War", *Communications of the ACM*, Vol. 30, No. 2, pp. 112–131.

[Brent87] Brent R.P., "A Linear Algorithm for Data Compression", *The Australian Computer Journal*, Vol. 19, No. 2, pp. 64–68.

[Card80] Card S.K., Moran T.P., Newell A., "The Keystroke-Level Model for User Performance Time with Interactive Systems", *Communications of the ACM*, Vol. 23, No. 7, pp. 396–410.

[Clare72] Clare A.C., Cook E.M., Lynch M.F., "The Identification of Variable-Length, Equifrequent Character Strings in a Natural Language Data Base", *Computer Journal*, Vol. 15, No. 3, pp. 259–262.

[Cleary84] Cleary J.G., Witten I.H., "A Comparison of Enumerative and Adaptive Codes", *IEEE Transactions on Information Theory*, Vol. 30, No. 2, pp. 306–315.

[Cleary84] Cleary J.G., Witten I.H., "Data Compression Using Adaptive Coding and Partial String Matching", *IEEE Transactions on Communications*, Vol. 32, No. 4, pp. 396–402.

[Cooper78] Cooper D., Lynch M.F., "Compression of Wiswesser Line Notations Using Variety Generation", *Journal of Chemical Information and Computer Science*, Vol. 19, No. 3, pp. 165–169.

[Cooper82] Cooper D., Lynch M.F., "Text Compression Using Variable to Fixed-Length Encodings", *Journal of the American Society for Information Science*, Vol. 33, No. 1, pp. 18–31.

[Cormack85] Cormack G.V., "Data Compression on a Database System", *Communications of the ACM*, Vol. 28, No. 12, pp. 1336–1342.

[Cormack87] Cormack G.V., Horspool R.N.S., "Data Compression Using Dynamic Markov Modelling", *Computer Journal*, Vol. 30, No. 1.

[DIGITAL78] Digital Equipment Corporation, "VAX11 Software Handbook", Digital Equipment Corporation, Maynard, Massachusetts, 1978.

[DIGITAL79] Digital Equipment Corporation, "VAX11/780 Hardware Handbook", Digital Equipment Corporation, Maynard, Massachusetts, 1979.

[DIGITAL81] Digital Equipment Corporation, "VAX Architecture Handbook", Digital Equipment Corporation, Maynard, Massachusetts, 1981.

[Dawes71] Dawes R.M., "A Case Study of Graduate Admissions: Application of Three Principles of Human Decision Making", *American Psychologist*, Vol. 26, No. 2, pp. 180–188.

[Dennett84] Dennett D.C., "Elbow Room", Oxford University Press, Walton Street, Oxford OX2 6DP, England, 1984.

[Dietterich85] Dietterich T., Michalski R.S., "Learning to Predict Sequences", Reports of the Intelligent Systems Group, File No. UIUCDCS-F-85-939, University of Illinois at Urbana-Champaign, Urbana, Illinois, 1985.

[Diffie76] Diffie W., Hellman M.E., "New Directions in Cryptography", *IEEE Transactions on Information Theory*, Vol. 22, No. 6, pp. 641–654.

[Eggers80] Eggers S.J., Shoshani A., "Efficient Access of Compressed Data", *Proceedings of Very Large Data Bases International Conference*, Vol. 6, pp. 205–211.

[Feller57] Feller W., "An Introduction to Probability Theory and its Applications", Wiley, 605 Third Avenue, New York, New York 10016, 1957.

[Fiala89] Fiala E.R., Greene D.H., "Data Compression with Finite Windows", *Communications of the ACM*, Vol. 32, No. 4, pp. 490–505.

[**Fischer84**] Fischer G., Lemke A., Schwab T., "Active Help Systems", *Readings on Cognitive Ergonomics — Mind and Computers*, Springer Verlag, New York, 1984, pp. 116–131.

[**Fowler65**] Fowler H.W., "Modern English Usage", Oxford University Press, Walton Street, Oxford OX2 6DP, England, 1965.

[**Freeman78**] Freeman M., Jacobs W.W., Levy L.S., "On the Construction of Interactive Systems", *Proceedings AFIPS National Computer Conference*, Vol. 47, pp. 555–562.

[**Gallager78**] Gallager R.G., "Variations on a Theme by Huffman", *IEEE Transactions on Information Theory*, Vol. 24, No. 6, pp. 668–674.

[**Gallant80**] Gallant J., Maier D., Storer J.A., "On Finding Minimal Length Superstrings", *Journal of Computer and System Sciences*, Vol. 20, pp. 50–58.

[**Garey79**] Garey M.R., Johnson D.S., "Computers and Intractability", W.H. Freeman and Company, New York, 1979.

[**Genrich81**] Genrich H.J., Lautenbach K., "System Modelling with High-Level Petri Nets", *Theoretical Computer Science*, Vol. 13, pp. 109–136.

[**Golomb66**] Golomb S.W., "Run-Length Encodings", Correspondence, *IEEE Transactions on Information Theory*, Vol. 12, pp. 399–401.

[**Golomb80**] Golomb S.W., "Sources Which Maximize the Choice of a Huffman Coding Tree", *Information and Control*, Vol. 45, pp. 263–272.

[**Gottlieb75**] Gottlieb D., Hagerth S.A., Lehot P.G.H., Rabinowitz H.S., "A Classification of Compression Methods and Their Usefulness for a Large Data Processing Centre", *AFIPS*, Vol. 44, pp. 453–458.

[**Guazzo80**] Guazzo M., "A General Minimum-Redundancy Source-Coding Algorithm", *IEEE Transactions on Information Theory*, Vol. 26, No. 1, pp. 15–25.

[**Guttag80**] Guttag J., "Notes on Type Abstraction (Version 2)", *IEEE Transactions on Software Engineering*, Vol. 6, No. 1, pp. 13–23.

[**Hamaker88**] Hamaker D.W., "Compress and Compact Discussed Further", Technical Correspondence, *Communications of the ACM*, Vol. 31, No. 9, pp. 1139–1140.

[**Hardy82**] Hardy I.T., "The Syntax of Interactive Command Languages: A Framework for Design", *Software Practice and Experience*, Vol. 12, No. 1, pp. 67–75.

[Havelock78] Havelock E.A., Hershbell J.P., "Communication Arts in the Ancient World", Hastings House Publishers Inc., New York 10016, 1978.

[Hazboun82] Hazboun K.A., Bassiouni M.A., "A Multi-Group Technique for Data Compression", *ACM-SIGMOD International Conference on Management of Data*, Vol. 1, pp. 284–292.

[Hellman77] Hellman M.E., "An Extension of the Shannon Theory Approach to Cryptography", *IEEE Transactions on Information Theory*, Vol. 23, No. 3, pp. 289–294.

[Helman82] Helman D.R., Langdon G.G., Martin G.N.N., Todd S.J.P., "Statistics Collection for Compression Coding with Randomizing Feature", *IBM Technical Disclosure Bulletin*, Vol. 24, No. 10, p. 4917.

[Helman88] Helman D.R., Langdon G.G., "Data Compression", *IEEE Potentials*, No. 2, pp. 25–28.

[Horspool87] Horspool R.N.S., Cormack G.V., "A Locally Adaptive Data Compression Scheme", Technical Correspondence, *Communications of the ACM*, Vol. 30, No. 9, pp. 792–794.

[Horspool88] Horspool R.N.S., "Review of Book "Data compression: Techniques and Applications, Hardware and Software Considerations"", Computing Reviews Review Number 8803–0148, *Computing Reviews*, Vol. 29, No. 3, p. 141.

[Huffman52] Huffman D.A., "A Method for the Construction of Minimum-Redundancy Codes", *Proceedings of the I.R.E.*, Vol. 40, No. 9, pp. 1098–1101.

[Jakobsson78] Jakobsson M., "Huffman Coding in Bit-Vector Compression", *Information Processing Letters*, Vol. 7, No. 6, pp. 304–307.

[Jakobsson82] Jakobsson M., "Evaluation of a Hierarchical Bit-Vector Compression Technique", *Information Processing Letters*, Vol. 14, No. 4, pp. 147–149.

[Johnsen80] Johnsen O., "On the Redundancy of Binary Huffman Codes", *IEEE Transactions on Information Theory*, Vol. 26, No. 2, pp. 220–222.

[Jones81] Jones C.B., "An Efficient Coding System for Long Source Sequences", *IEEE Transactions on Information Theory*, Vol. 27, No. 3, pp. 280–291.

[**Jones88**] Jones D.W., "Application of Splay Trees to Data Compression", *Communications of the ACM*, Vol. 31, No. 8, pp. 996–1007.

[**Karlin69**] Karlin S., "A First Course in Stochastic Processes", Academic Press, New York, 1969.

[**Katajainen86**] Katajainen J., Penttonen M., Teuhola J., "Syntax Directed Compression of Program Files", *Software Practice and Experience*, Vol. 16, No. 3, pp. 269–276.

[**Kauffman76**] Kauffman D., Johnson M., Knight G., "The Empirical Derivation of Equations for Predicting Subjective Textual Information", *Instructional Science*, Vol. 5, pp. 253–276.

[**Kernighan88**] Kernighan B.W., Ritchie D.M., "The C Programming Language", Prentice Hall Inc., Englewood Cliffs, NJ, 1988.

[**Kleinrock75**] Kleinrock L., "Queueing Systems", Wiley, 605 Third Avenue, New York, New York 10016, 1975.

[**Knuth73**] Knuth D.E., "Sorting and Searching", The Art of Computer Programming, Vol. 3, Addison-Wesley Publishing Company, Reading, Massachusetts, 1973.

[**Knuth79**] Knuth D.E., "Mathematical Typography", *Bulletin of the American Mathematical Society*, Vol. 1, No. 2, pp. 337–372.

[**Knuth81**] Knuth D.E., "Seminumerical Algorithms", The Art of Computer Programming, Vol. 2, Addison-Wesley Publishing Company, Reading, Massachusetts, 1981.

[**Knuth83**] Knuth D.E., "Literate Programming", Report No. STAN-CS-82-981, Stanford University, Stanford, CA 94305, 1983.

[**Knuth84**] Knuth D.E., "The TEXbook", Addison-Wesley Publishing Company, Reading, Massachusetts, 1984.

[**Knuth85**] Knuth D.E., "Dynamic Huffman Coding", *Journal Of Algorithms*, Vol. 6, pp. 163–180.

[**Kobayashi74**] Kobayashi H., Bahl L.R., "Image Data Compression by Predictive Coding I: Prediction Algorithms", *IBM Journal of Research and Development*, Vol. 18, No. 2, pp. 164–171.

[**Langdon79**] Langdon G.G., Rissanen J.J., "Method for Converting Counts to Coding Parameters", *IBM Technical Disclosure Bulletin*, Vol. 22, No. 7, pp. 2880–2882.

[**Langdon81**] Langdon G.G., Rissanen J.J., "Compression of Black-White Images with Arithmetic Coding", *IEEE Transactions on Communications*, Vol. 29, No. 6, pp. 858–867.

[**Langdon82**] Langdon G.G., Rissanen J.J., "A Simple General Binary Source Code", *IEEE Transactions on Information Theory*, Vol. 28, No. 5, pp. 800–803.

[**Langdon83**] Langdon G.G., Rissanen J.J., "A Double-Adaptive File Compression Algorithm", *IEEE Transactions on Communications*, Vol. 31, No. 11, pp. 1253–1255.

[**Langdon83**] Langdon G.G., "A Note on the Ziv-Lempel Model for Compressing Individual Sequences", *IEEE Transactions on Information Theory*, Vol. 29, No. 2, pp. 284–287.

[**Langdon84**] Langdon G.G., "An Introduction to Arithmetic Coding", *IBM Journal of Research and Development*, Vol. 28, No. 2, pp. 135–149.

[**Langdon84**] Langdon G.G., "On Parsing Versus Mixed-Order Model Structures for Data Compression", IBM Research Report RJ-4163 (46091) 1/18/84, IBM Research Laboratory, San Jose, CA 95193, 1984.

[**Langdon88**] Langdon G.G., Pennebaker W.B., Mitchell J.L., Arps R.B., Rissanen J.J., "A Tutorial on the Adaptive Q-Coder", IBM Research Report, RJ5736, June 1988, IBM Research Division, Yorktown Heights, New York, 1988.

[**Lawrence77**] Lawrence J.C., "A New Universal Coding Scheme for the Binary Memoryless Source", *IEEE Transactions on Information Theory*, Vol. 23, No. 4, pp. 466–472.

[**Lea78**] Lea R.M., "Text Compression with an Associative Parallel Processor", *Computer Journal*, Vol. 21, No. 1, pp. 45–56.

[**Lelewer87**] Lelewer D.A., Hirschberg D.S., "Data Compression", *Computing Surveys*, Vol. 19, No. 3, pp. 261–296.

[**Lempel86**] Lempel A., Ziv J., "Compression of Two-Dimensional Data", *IEEE Transactions on Information Theory*, Vol. 32, No. 1, pp. 2–8.

[**Lynch73**] Lynch M.F., "Compression of Bibliographic Files Using an Adaptation of Run-Length Coding", *Information Storage and Retrieval*, Vol. 9, pp. 207–214.

[**Martin83**] Martin G.N.N., Langdon G.G., Todd S.J.P., "Arithmetic Codes for Constrained Channels", *IBM Journal of Research and Development*, Vol. 27, No. 2, pp. 94–106.

[Mayfield72] Mayfield J., Parham R., Webber B., "Fundamentals of Senior Physics (Book 2)", Heinemann Educational Australia Pty. Ltd., 85 Abinger Street, Richmond 3121, Australia, 1972.

[Mayne75] Mayne A., James E.B., "Information Compression by Factorising Common Strings", *Computer Journal*, Vol. 18, No. 2, pp. 157–160.

[McCarthy73] McCarthy J.P., "Automatic File Compression", *International Computing Symposium 1973*, North-Holland Publishing Company, Amsterdam, 1974, pp. 511–516.

[McIntyre85] McIntyre D.R., Pechura M.A., "Data Compression Using Static Huffman Code-Decode Tables", *Communications of the ACM*, Vol. 28, No. 6, pp. 612–616.

[Mitchell87] Mitchell J.L., Pennebaker W.B., "Software Implementations of the Q-Coder", RC 12660 (57105) 4/22/87, IBM Thomas J. Watson Research Center, Yorktown Heights, New York, 10598, USA, 1987.

[Moffat87] Moffat A., "Predictive Text Compression Based upon the Future Rather than the Past", *Australian Computer Science Communications*, Vol. 9, pp. 254–261.

[Moffat88] Moffat A., "A Data Structure for Arithmetic Coding on Large Alphabets", *Australian Computer Science Communications*, Vol. 11, pp. 309–317.

[Moffat88] Moffat A., "A Note on the PPM Data Compression Technique", Research Report 88/7, Department of Computer Science, University of Melbourne, Parkville, Victoria 3052, Australia, 1988.

[Moffat89] Moffat A., "Word-based Text Compression", *Software Practice and Experience*, Vol. 19, No. 2, pp. 185–198.

[Morland83] Morland D.V., "Human Factors Guidelines for Terminal Interface Design", *Communications of the ACM*, Vol. 26, No. 7, pp. 484–494.

[Morrin76] Morrin T.H., "Chain-Link Compression of Arbitrary Black-White Images", *Computer Graphics and Image Processing*, Vol. 5, pp. 172–189.

[Murray-Smith87] Murray-Smith S., "Right Words: A Guide to English Usage in Australia", Viking, Ringwood, Victoria, Australia, 1987.

[Murray33] Murray J.A.H., "The Oxford English Dictionary", Clarendon Press, Oxford, England, 1933.

[Neumann44] Neumann J.V., Morgenstern O., "Theory of Games and Economic Behaviour", Wiley, 605 Third Avenue, New York, New York 10016, 1944.

[Nievergelt82] Nievergelt J., "Errors in Dialogue Design and How to Avoid Them", *Document Preparation Systems*, Elsevier North-Holland Inc., Amsterdam, 1982, pp. 265–274.

[Notley70] Notley M.G., "The Cumulative Recurrence Library", *Computer Journal*, Vol. 13, No. 1, pp. 14–19.

[Palme79] Palme J., "A Human-Computer Interface for Noncomputer Specialists", *Software Practice and Experience*, Vol. 9, pp. 741–747.

[Park88] Park S.K., Miller K.W., "Random Number Generators: Good Ones are Hard to Find", *Communications of the ACM*, Vol. 31, No. 10, pp. 1192–1201.

[Parnas85] Parnas D.L., "Software Aspects of Strategic Defense Systems", *American Scientist*, Vol. 73, No. 5, pp. 432–440.

[Pasco76] Pasco R., "Source Coding Algorithms for Fast Data Compression", Ph.D. Thesis, Stanford University, Stanford, CA 94305, 1976.

[Perl75] Perl Y., Garey M.R., Even S., "Efficient Generation of Optimal Prefix Code: Equiprobable Words Using Unequal Cost Letters", *Journal of the Association for Computing Machinery*, Vol. 22, No. 2, pp. 202–214.

[Peterson77] Peterson J.L., "Petri Nets", *Computing Surveys*, Vol. 9, No. 3, pp. 223–252.

[Peterson79] Peterson J.L., "Text Compression", *Byte*, Vol. 12, No. 4, pp. 106–118.

[Pratt80] Pratt W.K., Capitant P.J., Chen W., Hamilton E.R., Wallis R.H., "Combined Symbol Matching Facsimile Data Compression System", *Proceedings of the IEEE*, Vol. 68, No. 7, pp. 786–796.

[Raita87] Raita T., "An Automatic System for File Compression", *Computer Journal*, Vol. 30, No. 1, pp. 80–86.

[Reghbati81] Reghbati H.K., "An Overview of Data Compression Techniques", *IEEE Computer*, Vol. 14, No. 4, pp. 71–75.

[Reps84] Reps T., Teitelbaum T., "The Synthesizer Generator", *SIGPLAN Notices*, Vol. 19, No. 5, pp. 42–48.

[Rich78] Rich C., Shrobe H.E., "Initial Report on a LISP Programmer's Apprentice", *Interactive Programming Environments*, McGraw-Hill Inc., New York, 1984, pp. 443–463.

[Rich83] Rich E., "Users are Individuals: Individualizing User Models", *International Journal for Man-Machine Studies*, Vol. 18, pp. 199–214.

[Rissanen76] Rissanen J.J., "Generalized Kraft Inequality and Arithmetic Coding", *IBM Journal of Research and Development*, Vol. 20, pp. 198–203.

[Rissanen79] Rissanen J.J., Langdon G.G., "Arithmetic Coding", *IBM Journal of Research and Development*, Vol. 23, No. 2, pp. 149–162.

[Rissanen81] Rissanen J.J., Langdon G.G., "Universal Modeling and Coding", *IEEE Transactions on Information Theory*, Vol. 27, No. 1, pp. 12–23.

[Rissanen83] Rissanen J.J., "A Universal Data Compression System", *IEEE Transactions on Information Theory*, Vol. 29, No. 5, pp. 656–664.

[Ritchie78] Ritchie D.M., Thompson K., "The UNIX Time-Sharing System", *Bell System Technical Journal*, Vol. 57, No. 6, pp. 1905–1929.

[Roberts82] Roberts M.G., "Local Order Estimating Markovian Analysis for Noiseless Source Coding and Authorship Identification", Ph.D. Thesis, Stanford University, Stanford, CA 94305, 1982.

[Roberts83] Roberts T.L., Moran T.P., "The Evaluation of Text Editors: Methodology and Empirical Results", *Communications of the ACM*, Vol. 26, No. 4, pp. 265–283.

[Rodeh81] Rodeh M., Pratt V.R., Even S., "Linear Algorithms for Data Compression via String Matching", *Journal of the Association for Computing Machinery*, Vol. 28, No. 1, pp. 16–24.

[Rubin76] Rubin F., "Experiments in Text File Compression", *Communications of the ACM*, Vol. 19, No. 11, pp. 617–623.

[Rubin79] Rubin F., "Arithmetic Stream Coding Using Fixed Precision Registers", *IEEE Transactions on Information Theory*, Vol. 25, No. 6, pp. 672–675.

[Rubin79] Rubin F., "Cryptographic Aspects of Data Compression", *Cryptologia*, Vol. 3, No. 4, pp. 202–205.

[Ryabko87] Ryabko B.Y., "A Locally Adaptive Data Compression Scheme", Technical Correspondence, *Communications of the ACM*, Vol. 30, No. 9, p. 792.

[**Schachter51**] Schachter S., "Deviation, Rejection and Communication", *Journal of Abnormal Psychology*, Vol. 46, No. 2, pp. 190–207.

[**Schneiderman82**] Schneiderman B., "The Future of Interactive Systems and the Emergence of Direct Manipulation", *Behaviour and Information Technology*, Vol. 1, No. 3, pp. 237–256.

[**Schuegraf74**] Schuegraf E.J., Heaps H.S., "A Comparison of Algorithms for Data Base Compression by use of Fragments as Language Elements", *Information Storage and Retrieval*, Vol. 10, No. 9, pp. 309–319.

[**Schwartz63**] Schwartz E.S., "A Dictionary for Minimum Redundancy Encoding", *Journal of the Association for Computing Machinery*, Vol. 10, pp. 413–439.

[**Severance83**] Severance D.G., "A Practitioner's Guide to Data Base Compression", *Information Systems*, Vol. 8, No. 1, pp. 51–62.

[**Shannon48**] Shannon C.E., "A Mathematical Theory of Communication", *Bell System Technical Journal*, Vol. 27, No. 3, pp. 379–423.

[**Shannon49**] Shannon C.E., Weaver W., "The Mathematical Theory of Communication", The University Of Illinois Press, The University of Illinois, 1949.

[**Shannon49**] Shannon C.E., "Communications Theory of Secrecy Systems", *Bell System Technical Journal*, Vol. 28, pp. 656–715.

[**Shannon51**] Shannon C.E., "Prediction and Entropy of Printed English", *Bell System Technical Journal*, Vol. 30, No. 1, pp. 50–64.

[**Shapiro80**] Shapiro S.D., "Use of the Hough Transform for Image Data Compression", *Pattern Recognition*, Vol. 12, pp. 333–337.

[**Singmaster80**] Singmaster D., "Notes on Rubik's 'Magic Cube'", David Singmaster & Co., 66 Mount View Road, London, N4 4JR, UK., 1980.

[**Sleator85**] Sleator D.D., Tarjan R.E., "Self-Adjusting Binary Search Trees", *Journal of the Association for Computing Machinery*, Vol. 32, No. 3, pp. 652–686.

[**Sleator86**] Sleator D.D., Tarjan R.E., "Self-Adjusting Heaps", *SIAM Journal on Computing*, Vol. 15, No. 1, pp. 52–69.

[**Smith76**] Smith A.J., "A Queuing Network Model for the Effect of Data Compression on System Efficiency", *AFIPS*, Vol. 45, pp. 457–465.

[Steele83] Steele G.L., Woods D.R., Finkel R.A., Crispin M.R., Stallman R.M., Goodfellow G.S., "The Hacker's Dictionary", Harper and Row, 10 East 53rd Street, New York, NY 10022, 1983.

[Stone86] Stone R.G., "On the Choice of Grammar and Parser for the Compact Analytical Encoding of Programs", *Computer Journal*, Vol. 29, No. 4, pp. 307–314.

[Storer82] Storer J.A., Szymanski T.G., "Data Compression via Textual Substitution", *Journal of the Association for Computing Machinery*, Vol. 29, No. 4, pp. 928–951.

[Tanaka82] Tanaka H., Leon-Garcia A., "Efficient Run-Length Encodings", *IEEE Transactions on Information Theory*, Vol. 28, No. 6, pp. 880–890.

[Tanenbaum81] Tanenbaum A.S., "Computer Networks", Prentice Hall Inc., Englewood Cliffs, NJ, 1981.

[Tarjan83] Tarjan R.E., "Data Structures and Network Algorithms", Society for Industrial and Applied Mathematics, Philadelphia, Pennsylvania 19103, 1983.

[Tarjan85] Tarjan R.E., "Amortized Computational Complexity", *SIAM Journal on Algebraic and Discrete Methods*, Vol. 6, No. 2, pp. 306–318.

[Tarjan87] Tarjan R.E., "Algorithm Design", ACM 1986 Turing Award Lectures, *Communications of the ACM*, Vol. 30, No. 3, pp. 205–212.

[Teitelman72] Teitelman W., "Automated Programmering: The Programmer's Assistant", *Interactive Programming Environments*, McGraw-Hill Inc., New York, 1984, pp. 232–239.

[Teitelman72] Teitelman W., ""DO WHAT I MEAN": The Programmer's Assistant", *Computers and Automation*, Vol. 13, No. 4, pp. 8–11.

[Teitelman84] Teitelman W., "A Display-Oriented Programmer's Assistant", *Interactive Programming Environments*, McGraw-Hill Inc., New York, 1984, pp. 240–287.

[Teuhola78] Teuhola J., "A Compression Method for Clustered Bit-Vectors", *Information Processing Letters*, Vol. 7, No. 6, pp. 308–311.

[Tischer87] Tischer P., "A Modified Lempel-Ziv-Welsh Data Compression Scheme", *Australian Computer Science Communications*, Vol. 9, pp. 262–272.

[Todd85] Todd S.J.P., Langdon G.G., Rissanen J.J., "Parameter Reduction and Context Selection for Compression of Gray-Scale Images", *IBM Journal of Research and Development*, Vol. 29, No. 2, pp. 188–193.

[Tufte83] Tufte E.R., "The Visual Display of Quantitative Information", Graphics Press, Box 430, Cheshire, Connecticut 06410, 1983.

[Turner75] Turner L.F., "The On-Ground Compression of Satellite Data", *Computer Journal*, Vol. 18, No. 3, pp. 243–247.

[USDOD83] United States Department of Defense, "The Programming Language Ada Reference Manual", American National Standards Institute, Inc. ANSI/MIL-STD-1851A-1983, United States Department of Defense, Washington, D.C., USA 20301, 1983.

[Usher84] Usher M.J., "Information Theory for Information Technologists", Macmillan Publishers, London and Basingstoke, 1984.

[Wagner73] Wagner R.A., "Common Phrases and Minimum-Space Text Storage", *Communications of the ACM*, Vol. 16, No. 3, pp. 148–152.

[Wasserman82] Wasserman A.I., "Automated Tools in the Information System Development Environment", *Automated Tools For Information Systems Design*, Elsevier North-Holland Inc., Amsterdam, 1982, pp. 1–9.

[Waters82] Waters R.C., "The Programmer's Apprentice: Knowledge Based Program Editing", *Interactive Programming Environments*, McGraw-Hill Inc., New York, 1984, pp. 464–486.

[Weiss78] Weiss S.F., Vernor R.L., "A Word-Based Compression Technique for Text Files", *Journal of Library Automation*, Vol. 11, No. 2, pp. 97–105.

[Welch84] Welch T.A., "A Technique for High-Performance Data Compression", *IEEE Computer*, Vol. 17, No. 6, pp. 8–19.

[White67] White H.E., "Printed English Compression by Dictionary Encoding", *Proceedings of the IEEE*, Vol. 55, No. 3, pp. 390–396.

[Williams88] Williams R.N., "Dynamic-History Predictive Compression", *Information Systems*, Vol. 13, No. 1, pp. 129–140.

[Winston77] Winston P.H., "Artificial Intelligence", Addison-Wesley Publishing Company, Reading, Massachusetts, 1977.

[Wirth76] Wirth N., "Algorithms + Data Structures = Programs", Prentice Hall Inc., Englewood Cliffs, NJ, 1976.

[**Witten80**] Witten I.H., "Algorithms for Adaptive Linear Prediction", *Computer Journal*, Vol. 23, No. 1, pp. 78–84.

[**Witten82**] Witten I.H., "An Interactive Computer Terminal Interface Which Predicts User Entries", *Proceedings of the IEE Conference on Man-Machine Interaction, Manchester, England, July*, pp. 1–5.

[**Witten83**] Witten I.H., Cleary J.G., Darragh J.J., "The Reactive Keyboard: A New Technology for Text Entry", *Proceedings of the Canadian Information Processing Society National Conference, Ottawa, Ontario, May 1983*, pp. 151–156.

[**Witten86**] Witten I.H., Cleary J.G., "Foretelling the Future by Adaptive Modelling", *Abacus*, Vol. 3, No. 3, pp. 16–73.

[**Witten87**] Witten I.H., Neal R.M., Cleary J.G., "Arithmetic Coding for Data Compression", *Communications of the ACM*, Vol. 30, No. 6, pp. 520–540.

[**Witten88**] Witten I.H., Neal R.M., Cleary J.G., "Author's response", Technical Correspondence, *Communications of the ACM*, Vol. 31, No. 9, pp. 1140–1145.

[**Witten89**] Witten I.H., Bell T.C., "Review of paper "Data Compression"", Computing Reviews review number 8902–0069, *Computing Reviews*, Vol. 30, No. 2, pp. 101–102.

[**Wolff78**] Wolff J.G., "Recoding of Natural Language for Economy of Transmission or Storage", *Computer Journal*, Vol. 21, No. 1, pp. 42–44.

[**Wright87**] Wright P., "Spycatcher", William Heinemann Australia, 85 Abinger Street, Richmond, Victoria 3121, Australia, 1987.

[**Yannakoudakis82**] Yannakoudakis E.J., Goyal P., Huggill J.A., "The Generation and Use of Text Fragments for Data Compression", *Information Processing and Management*, Vol. 18, No. 1, pp. 15–21.

[**Zipf49**] Zipf G.K., "Human Behaviour and the Principle of Least Effort", Addison-Wesley Publishing Company, Reading, Massachusetts, 1949.

[**Ziv77**] Ziv J., Lempel A., "A Universal Algorithm for Sequential Data Compression", *IEEE Transactions on Information Theory*, Vol. 23, No. 3, pp. 337–343.

[**Ziv78**] Ziv J., Lempel A., "Compression of Individual Sequences via Variable-Rate Coding", *IEEE Transactions on Information Theory*, Vol. 24, No. 5, pp. 530–536.

[**Zunde81**] Zunde P., "Predictive Models in Information Systems", *Information Processing and Management*, Vol. 17, No. 2, pp. 103–111.

Index